Audio Mastering: The Artists

Audio Mastering: The Artists collects more than 20 interviews drawn from more than 60 hours of discussions with many of the world's leading mastering engineers. In these exclusive and often intimate interviews, engineers consider the audio mastering process as they, themselves, experience and shape it as the leading artists in their field. Each interview covers how engineers got started in the recording industry, what prompted them to pursue mastering, how they learned about the process, which tools and techniques they routinely use when they work, and a host of other particulars of their crafts. We also spoke with mix engineers and craftsmen responsible for some of the more iconic mastering tools now on the market to gain a broader perspective on their work.

This book is the first to provide such a comprehensive overview of the audio mastering process told from the point of view of the artists who engage in it. In so doing, it pulls the curtain back on a crucial, but seldom heard from, agency in record production at large.

Russ Hepworth-Sawyer started life once as a sound engineer and occasional producer, with over two decades' experience of all things audio. Russ is a member of the Audio Engineering Society and co-founder of the UK Mastering Section there. A former board member of the Music Producers Guild, Russ helped form their Mastering Group. Through MOTTOsound (www.mottosound.co.uk), Russ now works freelance in the industry as a mastering engineer, writer, and consultant. Russ currently lectures part time for York St. John University and has taught extensively in higher education at institutions including Leeds College of Music, London College of Music and Rose Bruford College. He contributes from time to time in magazines such as *MusicTech*, *Pro Sound News Europe*, and *Sound On Sound*, plus has written many titles for Focal Press and Routledge. Russ is also the co-founder of the Innovation in Music Conference series (www.musicinnovation.co.uk) and also the co-founder of the Perspectives on Music Production series of academic books.

Dr. Jay Hodgson is on the faculty at Western University, where he primarily teaches courses on songwriting and project paradigm record production. He is also one of two mastering engineers at MOTTOsound, a boutique audio services house situated in England. In the last few years, Dr. Hodgson's masters have twice been nominated for Juno Awards, topped Beatport's global techno and house charts, and he has contributed music to films recognized by the likes of *Rolling Stone* magazine and which screened at the United Nations General Assembly. He was awarded a Governor General's academic medal in 2006, primarily in recognition of his research on audio recording; and his second book, *Understanding Records* (2010), was recently acquired by the Reading Room of the Rock and Roll Hall of Fame. He has other books published, and forthcoming, from Oxford University Press, Bloomsbury, Continuum, Wilfrid Laurier University Press, Focal Press, and Routledge.

Perspectives on Music Production Series

Series Editors
Russ Hepworth-Sawyer
Jay Hodgson

Titles in the Series

Mixing Music
Edited by Russ Hepworth-Sawyer and Jay Hodgson

Audio Mastering: The Artists
Discussions from Pre-Production to Mastering
Edited by Russ Hepworth-Sawyer and Jay Hodgson

Audio Mastering:
The Artists

Discussions from Pre-Production
to Mastering

**Edited by Russ Hepworth-Sawyer
and Jay Hodgson**

Routledge
Taylor & Francis Group

NEW YORK AND LONDON

First published 2019
by Routledge
711 Third Avenue, New York, NY 10017

and by Routledge
2 Park Square, Milton Park, Abingdon, Oxon, OX14 4RN

Routledge is an imprint of the Taylor & Francis Group, an informa business

© 2019 Taylor & Francis

Library of Congress Cataloging-in-Publication Data
Names: Hepworth-Sawyer, Russ. | Hodgson, Jay.
Title: Audio mastering : the artists discussions from pre-production to mastering / edited by Russ Hepworth-Sawyer and Jay Hodgson.
Description: New York, NY : Routledge, 2019. | Series: Perspectives on music production series
Identifiers: LCCN 2018023211 | ISBN 9781138900066 (hardback : alk. paper) |
    ISBN 9781138900059 (pbk. : alk. paper) | ISBN 9781315707501 (ebook)
Subjects: LCSH: Mastering (Sound recordings) | Sound engineers—Interviews.
Classification: LCC ML3790 .A9 2019 | DDC 781.49—dc23
LC record available at https://lccn.loc.gov/2018023211

ISBN: 978-1-138-90006-6 (hbk)
ISBN: 978-1-138-90005-9 (pbk)
ISBN: 978-1-315-70750-1 (ebk)

Typeset in Times New Roman
by Apex CoVantage, LLC

This book is dedicated to all those mastering engineers who so generously shared their time and expertise with us. Obviously, without them and their work, this book wouldn't exist. A special mention is reserved for John Dent, who sadly passed away before he could see his interview here in print.

Contents

Acknowledgements

This book has taken a considerable amount of work to complete, not only by the "authors" but by the contributors, too. Behind the scenes, however, were an unsuspecting team who aided us in transcribing and editing. We'd like to acknowledge their efforts here.

Many of Jay's research assistants spent hours transcribing and editing these fantastic interviews. In this regard, we'd like to thank: Ted Peacock, Kara-Lis Coverdale, Jason Mercer, Mack Enns, Robert Celik, Amaal Bhaloo, Gauri Angrish, Zoe Tzernaes, John Petingalo, John "Chico" Barrett, and Mark Collins. Russ would like to thank Kallie Marie for her connections in NYC and further afield. Some interviews would never have happened without her.

There were colleagues who went "over and above" to help us out. An extra thanks to Patrick Dal Cin, who spent considerable hours parsing interviews for us, and Matt Shelvock, who was a rock in terms of organizing and proofing material for us early on. We should also thank Jeremy Graham, without whom this book would probably not have been released for some considerable decades. Pat, Matt, and Jeremy need a beer bought for them when you meet them—we'll be owing them for years!

Thank you,

Russ & Jay

Introduction

Both of this book's authors have dedicated their academic careers to writing about the production process, and from many different angles. We have published research written from our perspective as producers and engineers (*What Is Music Production*—Focal Press), as instructors and audio engineering pedagogues (*Practical Mastering*—Focal Press), and as analysts of record production and recorded musical communication at large (*Understanding Recordings*—Continuum; *Representing Sound*—Wilfrid Laurier Press). Moreover, we have spent considerable time working to establish music production itself as a viable field of academic study, namely, the field lately called "music production studies." The results of this effort can be read in the continuing series we edit for Routledge, *Perspectives on Music Production* (Routledge), and it can be witnessed first-hand in our biannual conference, *Innovation in Music* (www.musicinnovation.co.uk).

Over the past two decades, we have studied the mastering process intensely, academically, and creatively speaking, even as we continue to work professionally as mastering engineers ourselves. However, the musical and creative landscape we started our work in has changed dramatically. Mastering looks—and, more importantly, it *feels*—very different to us these days. Indeed, as the record industry changes, so mastering follows suit. One of our main goals in pulling this book of interviews together was to document how the industry's most successful mastering engineers have greeted and navigated this tumult and transformation. That said, the more things change, the more they stay the same . . . and the goals of the mastering process remain more or less the same, even as formats, tools, and aesthetic priorities and values change and change again. We both felt that our experience as academics and engineers—as "maker-thinkers," as it were—provided us with a fascinating and holistic insight into this ever-changing world. But we felt that we needed to seek out the views of some of the world's industry's elite to bolster our understandings. Anybody can write about change and transformation in the mastering industry from a "grassroots" level, of course. But how do the engineers working at the very top of the field encounter it?

In fact, very little at all has been written about audio mastering. A few textbooks exist, of course, and these books make excellent guides for beginners who are just starting the process of learning how to master musical material. But little has been written about audio mastering which considers the multitude of approaches to that craft which currently exist in the industry. This is surprising. Every record features an audio master, after all, even if its creators do not consciously undertake a mastering process for it. So why don't we hear much from mastering engineers themselves about what they do when they finish a record? The few texts which do address the mastering process tend to describe it more as a sequence of technical adjustments—that is, as a kind of psychoacoustic problem-solving exercise—rather than as the artistic and individual process it has become. If the interviews we've collected in this book have taught us anything about audio mastering, it's this: every mastering engineer works differently, and often using different tools, even if they pursue the exact same aesthetic goal, namely, producing the best record possible from the mixes they are given.

To address this broad lacuna, and to address the general quandaries we noted previously, we have endeavoured as much as possible to "get out of the way" in this book. We wanted to allow mastering engineers themselves to speak to us, and therethrough to you, about what it is they think they do when they set about to mastering a record and how they go about doing so. We also spoke with some mix engineers to get a sense of what they expect from the mastering process and how they accommodate that process in their work mixing records. And we spoke with some people who work on the tools that many mastering engineers now regularly use.

We asked each engineer the same set of questions, allowing for some improvization as the discussions evolved. The rationale for adopting this format was twofold. First, it allowed us to ensure that each engineer addressed similar topics, so we could get a broad spectrum of opinions about some of the more pressing topics which modern mastering engineers presently face. Second, as you will see, each engineer "dug in" to different questions, and seeing which topics engineers gravitated towards gave us a sense of which issues, in particular, seemed most relevant to each individual engineer. In our opinion, this gave us a fascinating insight into the particular contexts and cultural landscapes that each engineer worked within, which, in turn, allowed us to draw mastering not so much as a general—or, even, a generalizable—field of musical work but, rather, as a fully *lived* experience.

The questions we asked were as follows:

1. How did you become a mastering engineer?
2. How did you learn to master? Did you have any teachers?
3. What is mastering?
4. What is your personal opinion of mastering as a job? How does it compare to other jobs in the music industry?
5. What are the things you typically listen for—which aesthetic areas might you address—in a mix, regardless of genre?
6. Do you have a preferred signal chain? Will you take us through it, from input to output?

7. Do you use any "secret weapon" bits of gear?
8. What sorts of EQ moves do you tend to make?
9. How important is de-noising or editing in mastering? Can you give some examples of de-noising moves you might make, or have regularly made?
10. How, if at all, do you generally address stereo image? Do you engage in mid/side work? Do you use wideners?
11. Besides compression, what are some other ways you might achieve "loudness"?
12. Do you think the so-called "loudness wars" are over? If so, who won?
13. Do you master differently for different formats?
14. Do you have any experience with remastering? How is it different from frontline mastering?
15. How has mastering changed since you started?
16. Where do you see mastering heading in the future, as the record industry's troubles compound and grow?
17. What would you tell yourself if you could travel back in time to the day you started your mastering career?

We did our best to ask these questions to as many mastering engineers as we could talk to, working across both Europe and North America. We have divided the resulting interviews along this "continental divide," as it were, but we did so only because we could think of no other organizational conceit that made sense given the material. We conclude with a brief discussion amongst ourselves about the book, our experiences in relation to writing it, what we think it says about mastering and the mastering industry in general, and how our own experiences as mastering engineers ourselves relates to our findings. We offer this discussion in lieu of any sort of broad summary and conclusion, which we felt would only impose some overarching narrative onto the book that would distract from the interviews themselves. Indeed, if there is one underlying message we think these interviews collectively impart, it's this: mastering is a stubbornly unique artistic practice, undertaken by critical artists who often prefer to work alone, which emphatically resists standardization.

Not everyone wanted, or had the time, to talk to us. But we think what follows provides a broadly comprehensive view of the field as it exists today. We hope you find it as edifying and enjoyable to read as it was for us to write. Finally, we conclude this introduction with a note of thanks to those engineers who did speak with us. Their generosity in sharing their working methods, and their artistic worldviews, cannot be overemphasized. We are forever in their debts.

Sincerely,

Jay Hodgson and Russ Hepworth-Sawyer

Part One

North America

Gateway Mastering

Bob Ludwig and Adam Ayan

Established by mastering legend Bob Ludwig in 1992, Gateway Mastering remains one of the most successful independent mastering houses currently in operation. Ludwig's list of credits is legendary, including work on records by the Kronos Quartet, Steve Reich, Led Zeppelin, Jimi Hendrix, Phish, Megadeth, Metallica, Gloria Estefan, Nirvana, The Strokes, Queen, U2, Guns N' Roses, Tool, Simple Minds, Bryan Ferry, Tori Amos, Bonnie Raitt, Beck, David Bowie, Paul McCartney, Bruce Springsteen, the Bee Gees, Madonna, Supertramp, Will Ackerman, Pet Shop Boys, Radiohead, Elton John, Disney's "Frozen" soundtrack, and Daft Punk, to name a few. Ludwig is joined in his work at Gateway by 40-time Grammy winner Adam Ayan, another engineer whose stature in the field probably can't be overstated. We spoke briefly with Bob Ludwig by email before connecting by Skype with Adam Ayan.

BOB LUDWIG

Q: Would you go out on your own as an independent mastering engineer in today's environment?

If I was starting out, by all means it would be far preferable to be employed by a company. Being in a group of other engineers, one can learn so much from interacting with them. I can't imagine what learning mastering in a vacuum would be like. When I started with Phil Ramone at A&R Recording, I think I learned more from their superb maintenance staff than even the other engineers. Interacting with others is a font of knowledge from which to draw knowledge.

Q: How has mastering changed since you started? Does anything in particular seem lost or gained in today's professional climate?

Mastering started as a transfer process from tape mix to a reference lacquer. Then it evolved into doing more and more creative sound changes while doing the lacquer cutting, and finally it graduated to today's process of doing a two-track to two-track remix. When mastering was a part of the great studio systems, like Sony Music Studios, A&R Recording, The Hit

Factory, etc., one refined one's skill over the course of a career. With that gone, it is difficult to become a well-rounded engineer.

Q: Did you have any important teachers early on in your mastering career? Do you remember anything of particular significance that they taught you and that you pass on today?

Phil Ramone was my mentor. He taught me what a good sound was and how to best handle clients. He especially taught me how to remain cool under intense pressure. A&R Recording was the first independent recording studio to buy the Neumann computerized lathe. No one could figure it out, so Phil locked me and Aaron Baron into the room until we figured out how to make it work. I learned by comparing what I was doing with the union engineers at CBS, Capitol, Deutsche Grammophon, and RCA. Mastering has evolved so much from then. Me, Lee Hulko, Doug Sax, Bernie Grundman, and Steven Marcussen were pioneers in seeing how far the mastering envelope could be pushed.

Q: What, in mastering, do you think engineers very early on in their career should know?

Get the technical side of the job down so thoroughly that you don't have to think about it. Be able to master, listening to music with the right side of your brain, and keep the time where one has to use the left side of the brain to as little as possible.

ADAM AYAN

Adam Ayan is a Grammy Award-, five-time Latin Grammy Award-, and TEC Award-winning mastering engineer who works at Gateway Mastering Studios in Portland, ME. His credits are diverse and exemplary—including recordings for superstar artists such as Paul McCartney, Carrie Underwood, Katy Perry, Gwen Stefani, Lana del Rey, Shakira, Madonna, the Foo Fighters, the Rolling Stones, Kelly Clarkson, Barbra Streisand, Juan Luis Guerra, Rush, Rascal Flatts, Pearl Jam, The Animals, and Keith Urban, to name a few.

Adam has mastered several Grammy Award-winning recordings (40 to date!) and many No. 1 singles and albums and has well over 100 Gold, Platinum or Multi-Platinum projects to his credit. Many of the recording industry's best producers, engineers, and artists rely on Adam's ears and expertise to fully realize their artistic vision. You can find out more about Adam and his work at www.adamayan.com.

Q: How did you become a mastering engineer?

A: From the very beginning, I went to the University of Massachusetts at Lowell for their sound recording, technology, and music performance program.

Q: Were Alex Case and Will Moylan there at the time?

A: Will was, yes. Will started that program actually, in the '80s, well ahead of me. I went in the mid-'90s. Alex wasn't quite there yet. I think he came in as faculty a few years after I had graduated from the program, but of course I know both of them very well, and I stayed in touch with them. I know Alex well, and I know Will very well. Will was a professor of mine, and the director of the program when I was there. Both are great people, and in my opinion, they run the best program hands down for sound recording in the country.

I was a musician as a kid, and I really wanted to go to school for music performance. I wanted to be a rock star. I played in a lot of rock bands when I was a teenager, and when it came time to figure out what I was going to do with myself after finishing high school, I decided I really wanted to go to school for music performance. However, I had spent a little time in the recording studio as a performer and as a musician, and I thought, "Well, there's this great sound recording program not too far from me geographically at UMass, maybe I'll check that out and see what that's about."

Once I got into the recording program, my gears totally shifted to recording and audio, and then eventually to mastering specifically. It's interesting, you're mastering guys as well, and I'm sure you've felt over the years that not many people completely understand what it is that we do because we're such a niche part of the industry. When I was in school at UMass, I didn't know a lot about mastering, but I had asked Bob Ludwig to come and speak to our AES [Audio Engineering Society] Student Chapter during my senior year. I had asked a number of industry professionals to come and speak throughout the year, and every time I did, it was this great experience for me to get an overview of different potential careers in audio. I knew I really wanted to be in music production, and as I researched what mastering was, and then of course had Bob come down and speak to our student chapter of the AES, I realized that mastering was something I really wanted to do.

I then interned at a mastering studio in Boston and worked there for a little bit before getting a job as a production engineer here at Gateway in 1998. Of course, that was where, in both cases—in the studio in Boston and here—the rubber met the road. It was one thing to think maybe I wanted to be a mastering engineer, but then once I was having the experience in a mastering studio, I realized it was definitely what I wanted to do.

I don't know about yourself, but I know there are a number of people in the industry that didn't quite start out in mastering. They did tracking for a while, or mixing for a while, and found their way into mastering. That said, folks like Bob [Ludwig], Doug [Sax], Bernie Grundman, Ted Jensen, and a lot of those guys, started out in mastering early on, and that was really how I started. I did some multi-track recording, and I did some mixing, but very early on in my career I was working in a mastering studio. By the time I was about 25 years old, I was mastering records here at Gateway. So mastering has been the primary focus of my career since the beginning.

Q: How did you learn to master? Like you said, a lot of people aren't fully aware of what it is exactly a mastering engineer does. How did you learn that craft?

Well, I learned by being in the mastering studio and having the experience I gained in Boston. More specifically, I learned through experience here at Gateway. Within about a year of working as a production engineer, being a guy that did editing and cut production masters for Bob's projects, I heard he wanted to have an assistant in the room. I really wanted to be that guy. So that was really where I learned the art of mastering, as Bob's assistant for a couple of years.

I started by just hanging out with Bob in the room and seeing what he was doing. Bob gave me a little bit of a tutorial and information about what he was doing. This eventually led to me setting up for him in the morning before he got in, and then setting up and EQing (equalizing) tracks before he came in to start a session. That was really where the real learning of the art came into play for me, because I could set up the technical stuff I had under my belt at that point, and I could start EQing tracks.

He'd come in and say "Oh no, I would have done it this way, or that way," or "That's too much top, or too much bottom." Or whatever the case was. Eventually, though, he would arrive in the morning and say, "Oh, that's exactly how I would do it." So that work refined my ears and my skillset on the creative side of mastering to the point where we both realized that I had developed the aptitude and the temperament to be a mastering engineer. Bob said, "Why don't you start taking on and developing your own clientele?" So it was really through that assistant-mentorship role, specifically with Bob as my mentor, that I learned the art of mastering.

Q: What a way to learn!

I like to say it's like I got an Ivy League education in mastering!

Q: So what is mastering?

I like that question. It is a difficult question to answer, though, as you know. The way I describe it, in a nutshell, to somebody that asks me that question, is to say that mastering is a two-step process. The first "step" is the end of the creative process of making a recording, and that's of course what I do every day as a mastering engineer. The other "step" is prep for manufacturing and digital distribution. Because the creative side is what I do every day, and is the more interesting and exciting part of what we do, I would define that as the ability to make a recording sound as best as it can, and for it to translate well to the end listener, the fans of the band, and to whoever made the recording in the first place. My job is to make every recording that comes across my desk sound the best that it can and to help bring out the emotion and message of the music that the artist is trying to convey with their recording to the listener. Of course, that's what everyone is doing in the recording process, but we're specifically at the end of that creative process. We're the last step.

Q: Were there any sessions in particular that stand out as being really important learning experiences for you?

Collectively, each and every one of them was a significant learning experience, in one way or another. One that just jumped into mind when you asked the question was a greatest hits record for the band Megadeth that I was assisting Bob on. It jumps out for me because it was a greatest hits record, not a typical studio album. It was a series of snapshots in time, with tracks that may have had different record producers, different mixers, different engineers, etc. Of course, as a mastering engineer, our job is to make it all sound great but also make it sound cohesive. As you know, with a greatest hits album, or a compilation, what we work on is much more varied than usual. I remember that one session being a struggle. As an assistant setting up, starting to EQ some of the tracks for Bob, and working with him on it, finding a common ground for all of the tracks was hard, and we had to accept the fact that some of them were going to live as they were. That one [session] sort of jumps out at me in a number of different ways as one where I learned a number of things that are mindsets and toolsets that I carry with me still today.

I remember with this one track, I was having a really, really hard time getting it to sort of stand up, or be loud, or have any bite to it, and I kept going for compression and limiting. Then I had Bob looking over my shoulder and saying "Well, why don't we just try opening up the top end, it's really dull. I realized that having proper frequency balance was just as important for loudness as using a ton of limiting and compression. Those types of things, and that session in particular, jumped into my mind immediately as one where a lot of lessons that I learned in every session, every day, came together.

Q: What are the things that you typically listen for, or what sort of areas are you going to address, when you master, regardless of genre?

That's a good question. I'm one of these guys that jumps right in and starts working. Some mastering engineers may like to listen to a track down flat and see how it hits them before they start doing anything. I usually jump right in and start working. For every track, of course, the first thing I'm going to focus on is any corrective EQ, or any corrective measures, I have to take with the mix. In other words, if things jump out as being just plain wrong, those are the things I need to address first before I can delve into the craft or more creative part of what I'm doing. If there's too much bass response in the track overall, if it's boomy, the first thing I'm going to work on is getting that in line, fixing that, and making that work before I delve into anything else. So corrective EQ, or corrective measures, are the thing I look for first, which generally of course means that I start by creating a good overall balance, in terms of frequency response, that is appropriate for that recording. That would be the first thing I'd be looking for.

Of course most of what I do would fall under the umbrella of pop music, so vocals are really important. During the first or second listen I'd make sure that the vocals are hitting me just right. Are they present enough? Are they loud enough? Are they clear enough? Are they right where they should be? I work on very, very few instrumental recordings. Again most of the stuff that I do would fall under that pop music category in one way or another, whether it's pop music being commercially released, pop, country, all of that So the vocal is so important.

Q: Do you have a preferred signal chain that you work with when you're doing this?

Yeah, I do. For most mastering engineers, myself included, I think knowing your signal path really well is super important, and adding new things into it you do fairly cautiously, though I'm not hesitant to add new stuff into it. In terms of my signal path, what my preference has been recently, and it's changed in the past couple of years, but because virtually every mix is coming in digital now, and that's been the case for a long time, I'm finding that I'm staying all digital these days 99 times out of 100. If you asked me this question five years ago, I would say my preference was a hybrid signal path, even when given a digital mix. I would do a D/A conversion, I would use analog gear, and do a subsequent A/D conversion, use some digital plug-ins in-the-box, stuff, like that, but these days I'm finding my preferred signal path is all digital. I use both Pro Tools and Pyramix. Those are the two workstations I use every day. I use Pro Tools for playing back mixes, and for plugin processing, and then I use Pyramix for the other end, as the final step of my signal path, not including my monitoring of course, but the final step in terms of where the processed signal goes. All of it will go into Pyramix, and all the final editing, and the final master export, happen in Pyramix.

Q: Will you take us through your conversion and monitoring?

A: Sure. These days I'm using a Horus convertor, which is made by Merging Technologies, the folks that make Pyramix. I think they make the best-sounding converter on the market now, and so the end of my chain is Pyramix to the Horus for D/A (Digital-to-Analog conversion), through an SPL console; I have a DMC console, and Duntech Sovereigns are my monitors of choice.

Q: So you said that you're mostly digital now. Do you have any bits of outboard gear that you're still using nonetheless, somewhat regularly?

A: Not really, no. I do have a couple of TC devices that occasionally I'll go into for various things. And I have some multifunctional devices, like the system 6000 or, believe it or not, the Finalizer. Occasionally I'll go into one of those devices. In both cases, there are just a handful of algorithms or setups that I like in the devices, but it's becoming even less and less common for me to use them. I'm staying in the box almost entirely now. It's almost unbelievable that that's become the case, and I feel like I'm getting better results that way.

Q: Me too.

A: Yeah, isn't that interesting? For many, many years, I felt like using analog was a really important part of my signal path, and I challenged myself to do more in the box a few years ago when I had a couple of very intense, very time-intensive projects. It was like, "So, how am I going to get through these in a really efficient manner?" And I was like, "Let me see if I could do them all in the box." Of course, if I couldn't I would have dealt with it. But I realized at that point that everything in the box had become so good, and subsequent to those projects I found that anytime I was going out of the box I was getting lesser results than I used to. I think that some folks still really like analog gear, and I think there's still some mythology around it, but I'm definitely a convert to digital.

Q: Getting a little bit more into the nitty gritty, I wanted to talk about de-noising and editing. Do you do either often, and how much is either de-noising or editing actually a part of your practice?

Let's start with de-noising. I think it's something that happens every day, to one extent or another. For me, de-noising on almost a daily basis just means removing some clicks and pops and that sort of thing. Occasionally you get something with a lot of analog hiss, and maybe I'll go after that if it's bothersome. My general rule of thumb with de-noising is that if there is something that is not a musical component of the recording, and it takes me out of the moment when I'm listening to it, then I need to address it. That's my threshold of whether or not I need to de-noise something, and of course everyone's going to have a different threshold.

Of course, with everyone doing everything in the box prior to mastering these days, you occasionally come across some digital clicks and ticks that are probably byproducts of edits that were done in tracking or mixing. So I'll take those out. Whether it's a drummer putting their sticks down, or a guitar player moving in their seat in a quiet spot, I'll go after those. But again, my threshold with de-noising is always, "Does it take me out of the musical moment at all?" If so, I'm going to address it, and that happens on a daily basis.

In terms of editing, for me in the pop world, it's mostly topping and tailing—cleaning up the beginnings of songs, cleaning up the endings, doing fade-outs if the client hasn't done them. Transitions from song to song are still super important to me, and I spend a lot of time working on them when I'm doing an album or an EP project. You know, how do we flow from one song into another? I think that really makes a huge difference if you do that in the right musical way. So those are the two main things I'm doing in editing.

There's always the odd musical edit that needs to happen in a song. You may get a note from the producer, or the artist, saying, "Oh, we decided we need to cut the intro, and we didn't do it before we sent it to you. Can you do that?" So I'm doing those kinds of things as well. I'm also not afraid of editing between mixes if I'm given more than one master mix Often there are mixers that will provide me with a master mix and a couple of their variations of "vocal-ups" and if I'm given those, and I'm almost always given free rein, I will sometimes cut a vocal-up in, or a vocal-down in, somewhere to make the overall vocal level more smooth throughout a song. I don't hesitate to do that either. I think that sort of thing encapsulates the kind of editing that I'm doing on a daily basis.

Q: Do you address stereo image, and do you work mid/side at all?

I do. I address stereo image a little bit. I try to make things a little wider, when it's appropriate. It's not always necessary; it's not a de facto, default thing I always do, and I don't do it for everything. I often find, and I'm sure you do as well, that EQing and getting the frequency balance just right will often make the stereo image wider in itself. But I'll go after some stereo widening if I feel it's necessary, and if I feel it's beneficial to the recording. It's usually very subtle for me.

Q: Are there any tools in particular that you use, or techniques?

Yeah. The one tool that I use the most is actually not a super common one, because it only exists in one place. It's a stereo widener that SPL makes on their Vitalizer. It only runs on an AudioCube. I guess they have it licensed exclusively to Cube Tech International. I actually run this third computer that's almost strictly just for using this stereo widener [laughs]. There's a couple of other tools that are licensed only to that AudioCube, that I'll reach for occasionally as well, but that's my favourite one, because it's subtle, and stereo widening is something I do very, very carefully and subtly.

That brings me to mid/side processing, which is something I've never been a fan of. It's never worked for me, so I don't really use any mid/side processing. I've just never found it to be a useful tool. I've often found that with most of the tools I use for it, I've felt as though I could hear something happen, like a phase shift, that always ended up being a byproduct of the process that I didn't like. Maybe to further that notion, or to further explain why I haven't found it super useful: I do find that on a regular basis I get, what I think, are really good mixes, and I feel like mid/side is a tool that you can use creatively to fix a really problematic mix. So because I'm fortunate enough to get a lot of good mixes, maybe that's why I'm hearing more of the by-products, and less of the benefits, of using mid/side processing in my mastering So mid/side processing has never really been a tool that I've been one to use. I mention that only because that's what I feel when I've talked to engineers that do like to use it—I feel like they're using it to kind of save the day on a bad mix.

Q: So, we don't want to re-hash the "loudness wars," as I think it's becoming a tiresome topic in general, but we have to address it on some level. What sorts of uses might you have, or why might you reach for a compressor, other than to just "jack" the RMS level?

Compression is an interesting thing. You can use it to "jack" the RMS level or as a kind of "glue." I often find that compressors are imperative, especially with the kind of music that I work on. Compression's probably being used in tracking, in mixing, and in a lot of cases it needs to be used in mastering. So for me, it's a musical thing. It's kind of a "feel" thing, a glue. In some ways, it takes a mix that might feel, for lack of a better way of saying it, somewhat disconnected, and it makes it feel brought together a little bit, which I think is really important. It can be misused of course, but I think that rock & roll and pop almost couldn't really exist without some level of compression, right?

I think it's important to talk a little bit about the "loudness wars. I'm happy to talk about the "loudness wars," and compression. I think it's a bit of "beating a dead horse," but I think it's important, because I think there is misinformation for folks that really dislike the issue of loudness. They've kind of demonized compression. Compression is such an important tool in what we do, so occasionally, and this hasn't happened fortunately in a bit, but occasionally I'll have a client who says, "I don't care about loudness. I don't want to do this 'loudness wars' thing, it's so not musical, don't use any compression." And I'm like, "Are you sure about that?" And if I blindly follow what they tell me to do, and occasionally you get pushed into that corner, it doesn't end up being the same best result it could have been. So then, in the end, they realize that they had misunderstood what compression is about. So for me, it's about the musicality when you use compression; it's a glue, a "feel," that makes compression important. And then, of course, overall dynamics management. That said, I tend to use other tools for dealing with dynamics. Compression, for me, is more of a glue, a "feel" thing.

Q: That actually leads right into my next question. Besides compression, what are some other ways you might manage dynamics?

Mainly limiting, to get the RMS up and to manage the overall dynamic. Then, of course, there's making level moves, if they haven't been made in mixing, which sometimes they're not. That's kind of the power of the workstation, too. You can do all those things, you can automate them if you want, so you don't have to worry about them, and you can sort of set them and forget them once you get them right. That [i.e., making level moves] would be more my approach than strictly using compression, especially if I'm dealing with something that has really quiet parts and really loud parts. Again, in the pop music world, you want to maintain the dynamic, but you do need to rein it in again for the people that are listening to it in the car and in places where that's not going to translate, like a great classical record would on a hi-fi system or something.

Q: We've found that some mastering engineers will deliver different masters for different formats. Others will just say, "Here's the master, do what you will with it." Do you master differently for different consumer formats?

For me, 99% of the time, the audio or the mastering is exactly the same, and then it's about deliverables. Of course, for "Mastered for iTunes," we deliver 24 bit files because that's what Apple wants, and that's what yields the best result in terms of an AAC encode. For overall digital distribution, aside from iTunes or high-res, we'll deliver 44.1 kHz/16 bit. But the creative part of it really doesn't change. It's more about what the deliverable is.

When I master something, I master it at the highest resolution that it is given to me at, so I always have a high-res version to use for one thing or another, and then I just pass out different deliverables. I find that there's the 1% that might be a little different—in other words, where the client is really into doing a louder RMS version for most mass consumer formats, and then a less limited, quieter, more dynamic version for, say, vinyl, or audiophile type formats. And that's a very specific client that asks for that. It doesn't happen very often.

Q: When you're delivering for vinyl, are you delivering something like an interleaved stereo file that someone's going to cut, or are you delivering the premaster that is cut itself?

No, we're delivering the stereo interleaved file, basically at the highest resolution possible; a set of .wav files, one per side of the record. We don't cut lacquers here anymore; we haven't in a long, long time; we really don't want to [laughs]. There are a couple of people that we suggest our clients use for cutting, because the way we look at it is: we master the thing, once the clients approved it; they love everything about it, regardless of what format it's going to end up on; and the idea is, we give the highest resolution .wav files to somebody to cut, somebody that is very good at cutting and also who understands that the creative part of the process is over. It's like what mastering was 50 years ago; just translate this approved master onto a disc, with its inherent flaws and limitations of the disc, and do the best job. As simple as that sounds, I've found that there are only a handful of guys in the country that are doing that really, really well, and yielding the best results in terms of cutting their lacquers for subsequent manufacturing.

Q: Do you have any experience remastering?

Yes, quite a bit.

Q: In your experience, what are the typical reasons for remastering? And how is it different from what you might call frontline mastering?

There is a huge difference For your first question, the reasons for remastering in my mind are always just about better sonics. The idea is, how can we make this thing that's already lived out there in the world, that people already love, how can we make it sound better? Is there a possibility of making it sound better? Of course, that can go hand in hand with technological advancement, as well as change in consumer formats. So in my experience, remastering has always been about: how can we make this thing sound better? Then, of course, if the artist is involved, or the record producer, they may have some notion as to how they think it can be better. Do they really just want it to be a higher resolution, and a clearer picture of what they've already done? In other words, going back to the original flat tapes: oftentimes, I find that if I go back to an original tape, and I choose the proper playback device in my studio—we have a number of different sets of playback electronics and types of playback machines, each that sounds different from each other—if we choose the right playback machine, the right signal path, oftentimes, right away, I can already tell that it's better than the original release. It probably, of course, still needs to be mastered—it wouldn't be a flat transfer—but you immediately start hearing those things that make you feel like you're already yielding a better result. That's probably a combination of the better signal path and better technology that exists in 2015 than it did in whatever year [the record you're remastering] was originally released. I think I've been fortunate in the sense that all the remastering I've done I feel has been done for the right reason, which is to enhance the sonics of the recording. Like I said, sometimes the ideal will be, somebody, the record producer, or the artist, or whoever, they already loved the way it sounded, so you do your best to make it sound like that, but better. The stereo image sounds wider, and the whole thing sounds deeper front to back. Overall, you're getting closer to what was on the tape when they mixed it, than what they could have got 30 years ago.

Q: It's interesting. It almost sounds like that's where mastering started, with that transfer tradition. We have this mix that works on this one format, and now we have to transfer it to another format so it sounds great on that format.

Yeah, and of course that's sort of the start of that remastering process. And then doing what feels right, guided by the artist, or whoever may be involved in the remastering, as to what they want the remastering to be. Sometimes, somebody will say, "I made this record 30 years ago, but I want to make it sound like it came out today." Which to them might mean making it sound aggressive, punchy, and loud like a 2015 record, when the record was made in 1985. Whereas another artist might say, "Oh no, don't do that. Just make it sound better than it did before." The point being that I feel like in all of the instances I've been involved in with remastering, it's been benevolent. It's been about making this thing sound better, and then maybe being able to also put it out on all these newer consumer formats, whether that was SACD 10 years ago, or high resolution digital downloads. So that, I think, answers the first part of your question.

As to second part: number one, when you're remastering something, you always have to do your homework because you always want to beat it. You want to make it better. That's always my point in remastering something. We are going to make this a better musical experience than it was when it first came out. You want to make sure you don't make it worse, or you don't miss the mark in terms of making it better, so you always do your homework. Unlike frontline mastering, there's always something to refer to that you need to check against, and I'm doing that throughout any remastering; I'm always checking against something else. Now that we're in 2015, that something else could be two or three different iterations of something that's been remastered a couple of times, and you're always wanting to make sure that you're making it better. Again, this comes back to whoever (from the client's side) is involved in the creative decisions with you, what they're looking for from the remastering—always keeping in mind that, a lot of times, the remastering that we're doing here is of very famous records, and records that people know really well, and which were important records for the artist, and were important records to a lot of their fans. You want to respect that, to a certain degree. You don't want to give them something that's completely different, unless you're directed to do so. So I feel like with remastering, you always have to do your homework, and it's a different skillset than frontline mastering. But I feel like it informs your frontline mastering in a lot of ways. I think remastering records has taught me a lot of things about mastering over the years that I might not have learned if I was only doing frontline mastering.

Q: Can you elaborate on that?

There are some specific instances that come to mind. I'll first start by
doing my homework, and going back and listening to these recordings,
and asking myself, What did this sound like? What did millions of people
listen to, for years? What did the artist approve the first time around?
What's magical about that, and how does that sound relative to what's
on the mix master tape, before it was mastered? I mean that alone is an
education. Taking a flat mix tape and trying to make it sound like what
was mastered before, you learn so much about mastering by doing that.
It's a challenge to be, like, "OK, I'm listening to this song, I'm listening
to the master tape, and I'm listening to the original release of it. How
did they get from point A to point B?" So you start turning knobs, and
you try to get there, and you learn an awful lot about mastering that way.
There have been a number of instances in remastering where I have discovered tools that I didn't know before, that became part of my everyday
frontline mastering toolset. By that I mean specific equalizers, or specific
compressors. For example, I remastered a couple of records for Pearl Jam
about five years ago, two favourite records of mine, the Vs. album and
the Vitalogy album (the Vs. album was an especially favourite record of
mine as a kid). I discovered this Neumann EQ that we had sitting in our
equipment room that I never used for frontline mastering before because,
when I first started assisting Bob 16 years ago, we didn't use it in frontline mastering. It's a broader stroke EQ that just wouldn't always be useful in every frontline mastering job, and when I was working on this Pearl
Jam remaster, I realized that, sonically, it was the key to the EQ of this
album, and how it was originally done. Then that became a tool of mine
that I would go back to for frontline mastering when it was appropriate.
I think that's just one example of the kind of the thing you can learn in
doing remastering. But really, I think what it comes down to, with that
doing your homework part, is trying to find out how did they originally
get from point A to point B. From the original mix master tape, how did
they get to what ended up getting released, and how can I imagine that
that could be even be better? So, I think that challenge is so different than
frontline mastering, and in so many ways makes your frontline mastering
chops better.

Q: How has mastering changed, if at all, since you started?

Well, I think that it's changed. The most dramatic changes I've seen
are two changes, and one of them is that "beating of the dead horse" of
loudness. The expectation of how loud something should be in 2015 is
so dramatically different than it was in the late '90s when I first started
out, and I don't even mind going on the record and saying this, but there
are things that in terms of loudness that are expectations today that just
would have gotten you fired 15 or 20 years ago, they would have been
unacceptable. I don't really want to hop on the "loudness wars" thing
too much, but let me also tell you what my philosophy on loudness is,

because I think it's important to me at least, and I hope you'll find it interesting. I know I'm complaining about loudness a little bit right now because we're having this great conversation, but I don't complain about loudness that much because it's a reality. That's why I think it's "beating a dead horse" to complain about it too much, because it is a reality and the genie is out of the lamp, and it's not going to change that dramatically. Every time we think it's going the other way, it doesn't, in fact it goes farther into the realm of loudness, that's been my experience. So my philosophy, although that example was a negative statement, but the positive side to that is that my philosophy is that I know my clients are looking for a certain amount of loudness, and I know that's very important to them. I also know that if I don't do it that I won't get the gig and they won't work with me, and that's just a fact. So I know that they want that, but the challenge that's posed to me is, how I make loudness and how I make it the most musical, and I take that as a real serious challenge. How can I make my clients projects as loud as they want them to be but do it in the most musical way possible? Even you as a mastering engineer know, that's not the easiest thing to do because that takes a lot of finesse and hard work. I mean, it'd be one thing if you just said, well, I can just push it through more compressors, or distort the thing more or whatever, and do these heavy-handed approaches that get loudness in a very unmusical way, but that's not what I'm in the business to do, and that's not what I want to do with recorded music. So I take it as a very serious challenge to attain the kind of loudness that's expected today in 2015, and has been expected for a number of years, but to do it in the most musical way possible, and I really do try to do that. So, there's that aspect of it, and I think that's my biggest gripe with "the loudness wars," because I do feel like everyone can complain about individual recordings, and there are some individual recordings out there that are so exceedingly loud, and so distastefully loud, and I'm not going to say that it was the mastering engineer, or the mix engineer, or the artist, but who knows what the situation is, but there are some insanely loud records. So we can complain about individual albums, and how they've suffered from loudness, but I think a bigger issue is that the overall perspective of loudness and what a recording sounds like has shifted. In other words, the kind of distortion that has become acceptable, and almost an expectation, is that thing that would have gotten us fired 15 or 20 years ago. Now, it's become a thing where I can hear that as distortion and as something I would have avoided at all costs 15 or 20 years ago, but I think that a lot of people that are listening to music now might hear that as what a record sounds like, because so many of them sound that way. Maybe I'm going too far down the road of loudness, but I think that that's an interesting theory that I have, and it's kind of an educated theory based on just comments and things I've heard about recorded music for the past decade, where I feel like perspective has shifted to where there is a certain level of missing dynamics, and a certain level of distortion that's become almost an expectation.

Q: We interviewed Barry Grint, and he was talking about loudness, and his work on some Oasis records, and he said they wanted it as loud as could be, but he saw it as a musical thing, almost the way you might talk about processing a guitar for tone. His complaint was that it was just overdone now, sort of like what you've just said. Now everybody's just using the same guitar tone, as it were.

Yeah, it's just really interesting to see how that's all shifted.

Q: Where do you see mastering heading in the future?

I think I'm really optimistic that mastering is going to be just as important going into the future as it is now, and the reason is that we're such a creative part of the process. I still have so many clients (and by clients I mean mix engineers) that say specifically that they don't want to master things. So on the pessimistic side you could say, well, the industry has shrunk quite a bit. The recording industry has shrunk and budgets have become smaller, and that means corners are being cut things are being done more efficiently and more cost effectively. I think what that's meant for mastering so far is that there are occasions where either mastering is forgone, or it's being asked of somebody earlier in the process. Most people that I know that do mixing don't want to do any mastering at all because it is such a different discipline, just like I don't want to be asked to do mixing in a mastering session.

So I'm optimistic that we're going to still always be needed for our creative input. I've always said that we're not in the business of mastering for CD, we're not in the business of mastering for iTunes—we're in the business of mastering music, and it's our job and our duty to master music to sound as best as it can, for whatever formats people are listening to it on. So I think that shields us from what might be perceived as the negative outlook on the future, but I also have this really optimistic and positive outlook on the future of recorded music if we can get over this hump of demonizing streaming as a means of consumer delivery, because I find an awful lot of hope in streaming for the recording business, and I see some very technical parts of streaming that offer hope in terms of overall fidelity of the recordings that we're making.

Q: Can you elaborate a little bit?

Sure. The hope of the recording industry as a whole is that streaming is what people want, and we just need to find a way to give it them, and for them to find value in it where they're willing to pay for it. I think that opens the floodgates for a lot of subscriptions if we can get over that hump, and that's from people with different minds other than myself, people that understand the business side better than me.

On the fidelity side, there are two big things with fidelity that streaming could offer that would make some changes, if they were done right. One of them, as of right now, is that a lot of the streaming services deal with the loudness issue a little bit differently than FM radio does, or actually the polar opposite way of FM radio in some cases. So where FM radio would try to squash everything to keep a playlist at a consistent level from song to song, and we understand what the negative sonic aspects are of that, with streaming, some of the services will do the opposite. They'll look at it and say, well, what's the loudest thing, what's the quietest thing, and place a loudness value on everything, and then say, well, the quietest thing is down here at this level, and now we're going to shift everything else, using gain, down to that level. So as opposed to pushing everything up, they're shifting the louder stuff down, and in doing so, I think that exposes the negative sonic aspects of hyper-compression and this loudness thing we were just talking about, because you start to realize you're not getting the benefit of the "louder always sounds better," which is such a true phenomenon. So when you're comparing something that's been made hyper-loud, which you turn down to the same level as something that's not hyper-compressed and is a little more dynamic, you realize that the less compressed, more dynamic thing actually sounds better, and the hyper-loud thing sounds shallow and not as exciting, in an "apple to apple" comparison like that. So that's a hope I have in terms of fidelity. The other one is kind of esoteric, but I'll share it with you, because I think it's kind of interesting.

The subscription model completely, (and this is why it's been so hard to "get over the hump"), it completely changes the way the record business has operated since the dawn of the record business, on the business side of things. If we were to just talk about loudness again for a second: if you go out and buy a CD, or you buy an album at the iTunes music store, it's kind of a point of purchase, you buy it once, and whether or not you even listen to it doesn't really matter from a business perspective, at least for the sale of that recording. If you never listen to it, it doesn't really matter. But say you do listen to it, and I'm sure you've heard from folks that have listened to really hyper-compressed records, one thing they often say is that, "Oh, it gets really fatiguing at some point, and I don't want to listen to it anymore." So they turn it off.

Well, the streaming model is the complete opposite; the more time somebody listens, the more financially beneficial it's going to be. We could argue or debate and talk all day long about what that rate should be back to the artist and to everybody involved, and it should be as high as possible. I'm a huge proponent for artists being paid properly and placing value in what they've done. All that aside, what it really comes down to is that the more somebody listens to something, the more you're going to get paid. So it creates a financial incentive to make records that people do not want to turn off because of fatigue.

Q: That's a really good point.

Does that make sense?

Q: It makes total sense.

I've shared that with a few people, and they usually really get it. Some folks still can't get over the whole, "Streaming doesn't pay anything and it sucks." That needs to be fixed, of course, but if we're making records that are fatiguing, that people are going to turn off, it didn't matter in the CD world because you've already bought it, but it matters in the streaming world. So if you're making a record that is so loud and fatiguing, or the fidelity of it becomes bothersome and they don't want to listen to it, then you've cut off your revenue stream of that recording. However, if you're making something that people want to consistently go back to because they're not being turned off by its sonics, then you're giving yourself the opportunity to make a better living off of it. So I'm optimistic for streaming, and I'm optimistic for what it's going to mean for the fidelity of music. I'm optimistic for what it's going to mean for all of us in the record business that want to continue to work in this business and to thrive in it. I think it's going to happen.

Q: You've just made me a little more optimistic! Last question, and this one may seem like a throwaway, but we have gotten some very interesting answers from it. What would you tell yourself if you could go back to the very beginning of your career?

That's a really good question. The one thing I think I could tell myself, and that I feel as though I was fortunate enough to learn somewhat early on, but would have been great to know right at the very, very beginning: I'll share with you that for recreation and fitness, I'm a distance runner. I mostly run half-marathons, and I've run a couple of marathons, and I think that I've looked at my career as a marathon with "ebb and flow," and looking at that sort of big picture, long-term thing, it took me several years to have that sort of mindset. It's something that I think I would have benefited from knowing right from the beginning. It's a little abstract in terms of what I'm trying to say, but we get very emotionally invested in what we do, of course, and I was very, very "gung-ho" early on in my career, which was really beneficial to me in a lot of ways, but I think it's also good to look at a career as this long-term plan with some "ebb and flow," and to have a long-term mindset about everything that we do. Now, I don't think there's anything I did in the short term that hurt my career early on, but I think I probably stressed myself out more than I needed to early on in my career, which is what it comes down to. I know I'm kind of talking circles now, but in some ways it was a really positive thing because I was super, super driven, and I still am, but I think that looking at what we do as sort of a marathon and not as a sprint is important. Again, I feel fortunate that I discovered that early on. You can think of it that way, and I'm sure you have friends and colleagues that you can think of that have really burnt themselves out by doing too much of a sprint. This is the career that I want to have my entire life; this is what I want to do my entire professional life. So I'm glad I figured out early on that I need to treat it more like a marathon, and not a sprint.

Q: It reminds me of that old Zen adage, "The journey is the destination."

A: Right, yeah. That's a really great one. That's something I wish I had thought a little bit more about earlier on. I don't have any regrets, but to give you an answer, that kind of a notion is important.

Ellen Fitton

Ellen Fitton is a music industry veteran. She got her early education at SUNY Fredonia as a graduate of their Tonmeister program. This solid foundation and training has enabled her to follow the extremely versatile path her career has taken. Her experience is rooted in some of New York City's most historic recording studios. She began her engineering career at Right Track and Atlantic Studios, where she worked with artists like Foreigner,

Joe Cocker, and David Byrne, and learned the art of pop recording working with legendary producer Arif Mardin. She later spent many years at Sony Music Studios, where she was part of an elite group of engineers who were selected to handle classical recordings using prototype equipment in the early days of digital recording. There she worked extensively with artists like YoYo Ma, Isaac Stern, and Placido Domingo as well as numerous world-class orchestras in both the US and Europe. While there, she was nominated for a Grammy for her work on Sony's *100 Years Soundtrack*. Her career took another turn 12 years later, when she was asked to join the mastering team at Universal Records as one of their primary mastering and restoration engineers. Ellen is credited with exclusively remastering the entire series entitled *The Complete Motown Singles*, a 13-volume box set which entailed restoring every A and B side ever released on the famed Motown catalogue from the original masters. Her diligent efforts on this undertaking earned her a MOJO award in 2007. Ellen's career has taken yet another turn. She is currently working in live sound and production for theatre in places like Radio City Music Hall and Carnegie Hall, and is also mixing and mastering for network television broadcasts.

Q: How did you first become aware of a mastering process?

There are many people who start out saying, "I want to be a mastering engineer." I didn't start out that way. My career has evolved over time and taken many turns. I don't think initially I was too aware of mastering. Even once I was, my focus was actually on recording. I wanted to record; that's what I had a passion for. I really wanted to record rock & roll, but after a while there was very little rock & roll happening in NYC. While I was trying to figure out what was next for me, I was approached by a friend about doing classical work. As a classically trained musician, I was intrigued by that. I hadn't really considered that as an option before. I ended up taking the job doing that for almost 12 years. It involved location recording, editing, and some reissue work. Sadly, the bottom sort of fell out of the classical market, too, but what was still thriving at that time were reissues. So I focused on reissues and restoration and sort of came into mastering from that angle.

There are two areas that are considered mastering: frontline and reissue. When people refer to frontline, that's new material coming out through current studio work/recordings. Then there is "remastering," which is all about taking the old material and refurbishing it. There are similarities and differences to the process for each. But the goal is the same: to prep the mixed track/album for presentation to the outside world. In whatever format that may be.

Q: Can you talk a little bit about how the remastering process differs from what you called frontline mastering?

It's a bit of night and day, really. With "frontline" mastering, people are presenting you with something that's been made in recent history. More often than not these days it's all digital, and it's in what you might want to call "pristine condition." They're coming to you with something that is a viable thing, and then you're putting the icing on the cake, so to speak. It's a different mindset than reissues. There are all the loudness wars, which I don't particularly want to get into here. It's about, "This is the sound now." You have to make your thing sound better or more unique, given that these are the parameters/standards for today's material. The major thing that often needs addressing in frontline mastering is consistency. The way records are made today, it is very often more a series of "singles." An artist works with a few different producers and mixers during the course of making their album, often many different studios as well. So the mixes/tracks often have a different feel and sound. It's the mastering engineer's job to make them sound like they belong on the same album without destroying their unique qualities.

In terms of reissues, you're dealing with so many different time periods, you have to assess each thing that comes to you. At least this is how I approach it. As most reissue work comes from analog sources, the first issue you have to address is, "What is the condition of the master that you've received?" You could have the oxide peeling off, so how do we deal with that? Or there's a drop out in a key place, and how do we fix that? The tapes splices are all coming apart, or there's edge damage, and now you've got those things to deal with. The first thing you do is almost like healing the patient. You have to make the tape whole so you can actually work with it. In other cases, it's on a format that maybe you don't even have the right machine to play it back on, or it doesn't have tones on it, or it was done with an alignment standard that isn't used anymore, and you need to recreate the alignment standard. Initially, you just need to get what's presented to you into a playable, healthy format.

Q: Do you have an example of something that was done to "heal the patient," with regards to faulty tape or something like that?

There are all kinds of things. Like I said, oxide for one; there's a whole period of time where Ampex tape was not good. It did not hold up over time. You'd get the tape, and the oxide is literally sticking to the heads as you try and play it, so you'd have to bake it. The thing is, baking is only a short-term solution. You really need to bake it and get it transferred to a stable format. These days that's typically digital. There was something I had done with Motown. It was a really well-known single by Junior Walker that had been released and had been out there for years, on any number of Motown compilations, and we needed it for this 12-volume box set I was doing. There was a small but very noticeable drop out in the middle of the song. Since this was for the *Complete Motown* singles box set, a high-profile release, I turned to the producer and said, "Here we are trying to put out what we consider the benchmark for what all the singles on Motown were supposed to sound like, and we're going to put this out again, with the same drop out?" So I actually went back, and I listened to every other version that was out there, and they all had the same drop out! I said, "I don't want that to be the version that we put out for this monumental box set." We had to go back and look for out-takes, alternate mixes, trying to find that little piece. And even if we could find a better source, you never know if it's in good condition, does it match? Of course, if it matched, it would have been way too easy. You run up against issues like, is it the same mix? Is it the same level? EQ? Sometimes you try to intercut stuff, and [then you'd realize] they did different mixes, or they panned stuff different. You'd hear this insert go by, and you'd be saying, "Oh, that's not going to work." So, you'd have to figure out how to adjust that. In some very rare cases, we'd go back and try and hunt it down on multi-track. Just to remix a few notes! Multi-track back then isn't what multi-track is now. Multi-track then might have been three channels, so maybe I can't adjust it as easily as you want because of the track layout, or maybe it has reverb on it that the master didn't—there are challenges like that that you're presented with, and you need to be patient and resourceful to get through it.

Then you'd probably approach the one place where reissue and frontline are the same. You're trying to make it sound sonically as good as it possibly can. Is it as bright, crisp, and clear as you want it? Is there enough bass? Where does the vocal sit? There's all that kind of thing, where you want it to sound good coming out of the speakers, you want it to be pretty and enjoyable and have air and low end—but again, keeping in mind this was 1962 or this was 1970, and maybe the vocals weren't in your face, as that wasn't the style back then. Certainly analog distortion was par for the course, so the thing I would always try to focus on was making it sound as good as it could, yet as if I were in that time period. It's almost a question of preserving the approach to the production values of the time period.

Today we do things in a particular way that they couldn't do back then. The goal is to make it sound as pristine as I think they would have hoped that it could sound like, when it was originally done, if they had the tools to work with that I do.

Q: Is it common to try and preserve mix balances? A cynical point of view on remastering holds that people take these old masters and just crush them for modern listenerships . . .

I don't know what the percentage is. I can't say, unless we all sat in a room, had drinks and discussed our philosophies, but I do get that question a lot. I've heard, and people comment when I listen to my peers' work versus stuff I've done, "Oh yeah, their stuff is really much louder than mine." It's a matter of taste and there is a sense that some people really want it [loud]. They're probably not as fixated on preserving the time period as I am. They're about, "Well, we want it to sound like it was done today, because that's who we are competing with on the radio." Having said that, I don't necessarily think that some Motown track is going to come up on the radio right after a big hip-hop track. But on people's iPods, I guess that statement is true, so there's an argument to be made there. You could have somebody who is listening to a wide range of styles and different types of music, and when [an older track] comes up on their iPod or iTunes, they don't want that track to be lower in level than the newer tracks. I try and find a happy medium; that's the way I approach it. Of course, these days, people can set their devices to automatically level everything. I'm not a big fan of that, but I get the reasoning. Hopefully people don't have that on all the time when they listen. I think it starts to make everything sound like a relentless wall of sound. Music needs to breath, to ebb and flow.

Q: You've done work in a lot of different genres, and particularly in the jazz and classical world. The attitude or perception of the classical, and even the jazz, world is that records aren't mastered. Have you come across that at all?

Yes, and it's more so probably with classical, because it's a very purist approach to making a record. They're spending a lot more time upfront as they make that recording. They decide precisely where that mic is going to be and precisely what the balances are going to be, and depending again on whether they're recording to multi-track, or if they're going right onto two-track. In a lot of cases they're going right onto two-track. Especially in the old days, a 24-track recording of a classical performance just wasn't done. It was two-track, or maybe it was four-track. The whole idea of multi-track and overdubbing, all that stuff, is not how classical records get made. In place of overdubbing in the pop world, classical instead does editing. They'll do multiple takes, they'll not necessarily be start to finish every time, the producer is sitting at the session(s) with the score, and they're trying to determine whether the flute is sharp or flat in this particular measure, was everybody together in that measure? Then when the session is over, they're marking up the score, and saying things like, "I want this bar from this take, and this bar from that take." So everything is getting cut together. And there are often hundreds of edits, regardless of whether it is multi-track or two-track.

Q: Can you talk us through the process of how those notes are dealt with in the production process? For example, do those notes go to you eventually?

Here's where it gets into a little bit of a grey area, because if it's a two-track recording (more than half the time that's what it is), it's going to an editor after it's recorded. I was (and am) an editor, so you're cutting it all together. Once it's all cut together to their satisfaction, it's basically done. They don't, like you said, go onto discussing, "Well, now there's a mastering phase." They've done all their work upfront, in terms of getting it to sound the way they want it to sound. After you cut it together, you're sitting with the producer, and maybe adjusting some minor level things, or listening to how much spacing is between movements or pieces. It's a rare case when you are adding any additional processing, and if so, definitely not compression. Then they call it a day. You have to keep in mind with classical records, they are done on location. The ambience/reverb is part of the original recording. There is typically no digital reverb, or delay or other effects added. No compression, no auto tune . . .

Q: What about now, with the emergence of "Mastered for iTunes" as a format, I've noticed that a lot of "classical" labels like to release with the "Mastered for iTunes" stamp on the release. Could you comment on that at all?

The one thing I will say about classical is that they've, in a lot of ways, always embraced the changes in digital technology a little quicker than the pop genre. For example, they wanted to go to CD before anybody else. Because when they're making records and they have a piece of music that's 50 minutes long, a longer format serves their purpose. They also use a greater dynamic range in their recordings. As a result, they embraced digital first. It's part of the classical culture to make recordings as pure as they can. It's not just, "we want the mic in the perfect position" and "we want the cleanest shortest signal path that we can." Now, with Mastered for iTunes, it allows them to maintain the same higher sampling rates and better bit depths and all that. They moved on to that very quickly, because that's one less compromise in the process that they have to make.

To go back to the previous question, and to address the mastering for classical process, the example I gave you was two-track, which is a lot of what classical is. There are multi-track classical records that are made, but still, there's that editing process and there's a mix process, but not so much a mastering process. They will, however, do a very detailed mix phase. With jazz, and I don't have a ton of experience with jazz, but I have found there's a little more use of mastering in jazz. Guys like Wynton Marsalis and various others go through a mastering phase of the operation, so they're a little more of a hybrid of the classical and the pop techniques. You don't really see a lot of overdubbing in jazz either, but they'll do some along with editing, but maybe not to the extent of editing that classical does. Then they'll do some mastering, but not a ton of mastering, although there are phases to their process.

Q: The classical approach to recording means that a lot of it is on location. Given this, do any noise issues end up being addressed in editing, as opposed to if everything's being reduced to two-tracks on location? It seems like if you were given a two-track of a live rock recording, and you were doing de-noising, you would call that mastering.

Yes. As a classical engineer per se, you're wearing a few more "hats." Some people lean more towards the recording aspects, some people are more involved in the editing/mastering, but they all do it all to some degree. But yes, you're right, because you're on location, that presents a whole set of other issues; you always have to be careful. Aside from recording the artist, don't forget to record some room tone at the end of the day; not only do you need it to hook the movements together, but sometimes if there is some underlying air conditioning or street noise, or whatever, and you need to filter that out later, then you have a sample of it to use as a baseline. Again, this plays into classical being more willing to embrace certain technologies. For example, when Sonic Solutions first started, they came out with their NoNoise system. Suddenly, you were able to take out a lot of these underlying room noises and audience noise, which were things that could never be handled before. This also caused a big backlash, because when people first started using it, people were so enamoured with the notion that "Oh, I can suck all this out," that they overdid it. There then came an additional backlash that "Oh, NoNoise is terrible, and it's not part of the recording art form that we as classical engineers are doing." So, it came down to a really good, competent engineer, who had to find where the line is, then, between which noise do we leave in? Which do we take out? How much do we take out so it's not compromising the integrity of it? That was a whole other thing that was going on at that period in time.

Q: Is there a specific example that makes you think, is this noise, or is this not noise?

I don't know if off the top of my head I can pick a particular piece. I can go to generalizations like being able to take out tape hiss. Suddenly everybody was so enamoured with taking out tape hiss, but it's analog recording. If you take out all of the tape hiss, you're not doing that without consequence. You have to realize, people are okay with a little hiss. A little hiss is not a bad thing. It's when it's excessive, and it's hard to distinguish the programme material from the hiss, or it's distracting to the listener— that's when you really need to address it. It's the same with sources that were coming off lacquers, with all the pops and clicks and thumps. Well, some of that comes out, and you don't even notice it. Other times, you've got to watch when you take out those thumps that you're not taking out the fullness of the low end. You don't learn that overnight. It's something that, over time, you develop your own sense of where the line is.

Q: Do you think people are in any manner becoming less tolerant of noise, as there's very little surface noise on digital playback devices? I wonder if, to a certain extent in the historical recordings, the noise is almost a marker that, "This is an authentic historical recording," whereas something that is coming out today, if it had excessive noise, would it really be tolerated, even within the on-location recording? Is that a debate that still exists?

That's an interesting question, because if you listen to modern stuff, which for lack of a better word I'll keep saying "pop" recordings, even though I mean modern pop, hip-hop, dance, and so on, there's a lot of distortion out there. It boggles my mind that people are okay with how much distortion there is. Then on the flip side of that, are people hearing the hiss, and is it bothersome?

I think that there are a lot of untrained ears out there that are listening to our work. On one hand, I think people aren't paying attention because they're listening to distorted mp3s. On the other hand, their equipment has the ability to playback with a fidelity that it didn't used to, so I guess maybe they do notice the hiss a little more? Then you've got some artists that are adding surface noise to their mixes, so I don't know. I think it's a crazy mixed up world out there about what is acceptable. If you're talking about people who really are educated and understand, say, classical music, and that's what they're listening to, I think that they are okay with a certain amount of hiss. It's all about it not being intrusive.

Q: You do, and have done, a lot of work in different aspects of production. How would you say that it informs your work as a mastering engineer? If it does, how would you say that your awareness of these various stages informs your work as a mastering engineer?

I think that there's no such thing as too much education or too much experience. I guess as I've evolved from one thing to another, I've brought that knowledge and experience to the next thing. At times, that may have made me approach things a little bit differently compared with somebody who'd been doing strictly mastering for a long time. Maybe that approach was good, maybe it wasn't. I remember when I initially went from doing the pop recording to then getting into classical, I'd be on location and they're doing recordings, and I'd be asking questions, because initially I was the assistant. I'd ask things like, "Why are you doing it this way? Why aren't you doing it such and such a way?" I would sometimes get these sideways looks from the senior engineers who'd be like, "Oh, that's a 'pop' approach. We don't do that." I'd be like, "Well, why don't you overdub that section?" and they would say, "This is classical. We don't do that." As I followed it through to the editing process, then I better understood their process. There were other things along the way that I came to realize, "Okay, this is not the philosophy of how this genre works," but it didn't mean that there weren't places along the way where my experience in a different genre did help me get past some issue. In the remastering fix scenario I described earlier, my determination and, ultimately, my method for hunting down and creating a fix for that drop out was due in part to my previous pop/multi-track studio work. I knew what I could create if I had the right raw materials, and I also had the experience to manipulate those resources.

Q: Do you think that the route you take to mastering is important, in terms of how you approach it with regards to the production process itself?

Well, I think it affects the end result of your work methods. I'm not going to say that there's only one route to get to mastering. Mastering is like I said: it's the final stage, you're taking somebody else's vision, and you're trying to magnify all the good stuff and to put it over the top. Putting the icing on the cake, as I've said. You're putting your fingerprint on it, and that's a very unique thing, so you get there through your own process. I think having an understanding of every phase, recording and mixing, I can listen to something and it gives me insight into how the track in front of me got to this point. An understanding of the journey the music has been on helps me see what the destination should be.

Q: We find how people view mastering fascinating in general. We were talking with Ken Scott a while back. He was telling us that at Abbey Road, before you were allowed on the floor to do any audio engineering, or even touch a mic, you had to spend time in the cutting

room. We found it really fascinating that you had to spend time at the end of the line.

Yeah, I think it absolutely does help you if you understand the whole process. No matter which end you start from. That way, for example, if you're doing something on the record end (especially in those vinyl cutting days) and you're being asked to record something you know that you're never going to be able to get on the cutter—for example, if there is a huge low bass excursion—you might not be able to get that on disc. To know what's going to translate, and to know the kinds of things that they're looking at on the other end—I think that's great, I think that's part of making you a well-rounded engineer. If all you've ever done is mastering, and you have no idea how this material is recorded, I think that that would hamper your approach a little bit, to not understand why it's coming to you sounding the way it is. Maybe you don't understand why it sounds boxy, but it sounds boxy because it was recorded in a certain type of room. So, understanding what has caused it to sound a certain way I think would help you to know what tools to turn to [to] solve the issue. "Well, how do we open it up and make it sound like it was in a nicer room?" You would use a different tool to address that than, say, just EQ. So understanding what you're hearing and being able to identify what caused it, I think, is a very valid approach.

Q: Do you have a preferred signal chain? Is there any specific technology you like to use, or do you have a particular monitoring situation you swear by?

I would say that given my path within the industry, as I went primarily from one label-owned studio to another, I generally inherited a fair amount of what I had to work with. Maybe there was budget for some minor additions, often there wasn't, so it was a lot about learning to work with what I had. I was never in a situation where a room was built to my spec, as some are lucky enough to encounter.

I typically found a signal chain that worked with what I had. I can tell you over the years things that I really liked. I loved the DCS Converters, and the Sontec EQs, they were just fabulous. Maybe it was the DCS into the Sontec, which over the last few years was really my big thing. As opposed to other converters that I had, I just liked how the DCS converter sounded through the Sontec. Sometimes I processed "in the box" as well if there was some sound I was looking for but couldn't achieve on the way in.

TC Electronics makes this thing called the Finalizer, which I loved, and I know a lot of people "thumbed their nose" at it. Well, I thought it really worked great for all this Motown material I was doing. There is one thing about the Motown stuff, which I talked earlier about in terms of that period of time, that it really helped with. Nowadays the vocal is really upfront and in your face; it wasn't necessarily like that in the Motown days. I always wanted to pull it out just a little bit, because again, it's about the compromising of the old tradition with the new, and there was a setting on the Finalizer that I could always go to. It allowed me to just grab the vocal and pull it out just a little bit, to just do that thing I was always looking for. I didn't want the vocal crushing and in my face, but I wanted it set out a little bit in front of the band, a bit more than maybe the two-track, or the mono version that I had and was able to do.

Q: How would you go about doing that?

I can't tell you all the secrets, but like I said, there was a setting on the Finalizer that I liked for that; they actually stopped making them, and sadly I don't have one anymore.

As to the monitoring part of the question, there have been good monitors coming and going over the years, and you learn to adapt. I used what I had been handed in my room, and this is what I had to deal with. Not everybody was going to buy me the big fancy ten thousand-dollar monitors.

Q: It seems like the part of mastering that often doesn't get discussed is the fact that many mastering engineers tend to eventually compile their own gear for their own home setup. Was there something that you were looking for in particular, or was it more that you found this one piece of equipment that you really liked and built from there?

I would say it's the latter. The speakers were the first thing. I spent ages listening to different speakers and finding what I wanted that was in my price range. Then from there, I thought about what other little box or plugins would suit my needs, and yes, I use plugins at home, I'll admit it and probably be chastised for it.

I have these little Neumanns in my home mastering room now, which I love. Everybody's like, "Neumann doesn't make speakers, they make microphones?" Well no, they make speakers, and I love them. I also love my little IsoAcoustics [monitor isolators], which was one of these things I read about in *Mix Magazine*, and I'm thinking, "That can't possibly work." But I got myself a pair, and when those combined with the Neumann speakers, I've got a decent little setup here in my house now. It's not the same as having a very expensive place in the city in a nice room that's all custom acoustics, but it helps me do what I need to get done at home.

Q: A lot of the engineers we're talking to work in the box.

I would much rather have a pair of Sontecs at home, but I don't, and you can't buy them. For the kind of work I do at home, people only want to pay you $75 to master their single. So I use a plugin, that's what gets the job done. Again, you can "dog" different kinds of tools, but in the end it's how you use them. If you can use them in a good, tasteful, sonically pleasing way, then go for it.

Q: It's all about the product, right?

Yes. Look, I love Kraft Macaroni & Cheese. It's not gourmet, but I think it's the best mac and cheese out there; it's a staple in our house. So again, just like you're presented with a piece of music, and you've got to make it sound better, sometimes you're presented with a set of tools, in terms of the equipment, and you have to find a way to work with them. Maybe it's putting the EQ first, before the compression, or vice versa; or maybe it's about mixing up the chain, or about adding a little something from a plugin at the end, even though you've done this whole beautiful analog chain. I think you have to be open to the idea that there's multiple ways to approach it, and that there's not necessarily a fixed, traditional way to get to where you want to go. It's not going to be the same every time.

Q: What about your workstation?

I will say I was a diehard Sonic Solutions fan for many, many years. They eventually started to not keep up with certain aspects of what engineers needed the system to be able to do. At home now I have Pro Tools, and it's cost effective and great for certain things. I think it's a horrible editing platform—I still think Sonic is the best editing platform out there, but it's a terrible mixing platform. You can't burn CDs or do PQ on Pro Tools. Sonic was a little slow to adapt to going to formats like .wav files, and dragging and dropping different types of formats, and certain things about getting on and off the system got cumbersome. Still, if I ever have a really difficult edit to do, and it's not multi-track, I'll do it on Sonic. If I have NoNoise work to do, sometimes I'll go to Sonic, sometimes I'll go to iZotope; they have some good plugins [RX], and their stand-alone is really good, so I use that. I will say Sonic has caught up a bit in recent years, but sadly they waited too long and people moved on. Not just to Pro Tools but Sequoia and Pyramix, too.

Q: So you use a DAW almost more for a specific task?

Yeah, it's a little task based. I've tried Pyramix and SADiE. I've dabbled with all of them. Pyramix is a really powerful platform, but for me, I thought they made it overly complex. I have used them all, but Sonic (actually now they go by the name Soundblade) and Pro Tools are my go-to platforms.

Q: What about metering? Do you have a particular meter that you trust, or is it adaptable?

Not really for home. I tend to try and trust the metering within the workstation. I do have a USB Pre Box with metering on it, so I have sort of a backup. But I don't have a big fancy meter.

Q: Is there anything as a mastering engineer that you wish mix engineers knew in general about mastering, that might help make the process smoother?

Yes! Don't send it to us already squashed. That's really the key. Once you squash it, I can't un-squash it. For the pop material I get that a lot, because everybody's making reference mixes to send out before it's been mastered, to listen to on this system or that system. Again, it's the loudness thing.

They want it to be as loud as everything else that they're comparing it to, so they're squashing it. Look, it's okay if that's what you need to do to get your reference, to see whether you need to adjust your mix, but then take it off before you send it to me. Like I said, there's no backing away from it—it's overdone.

In the mastering environment, and I mean absolutely no offence to mix engineers, but sometimes we mastering engineers (and our systems) are a bit more sensitive to the distortion angle. Our environments strive to be, at least in the "golden years," pristine listening environments. The idea was always to come as close to as perfect a listening environment as you could get, compared to all the studios where the recording, mixing, and tracking was happening. Even though they strive for that, lots of places just aren't. So the idea originally for mastering was to come to a place where you can truly trust how it sounds and hear what you couldn't hear in other places. That's not so true anymore, just because there's not as many premier mastering houses, and so much more is being done at home and all of that, but our focus is much more honed-in on certain things. I'm just trying to say that I think sometimes we're more sensitive to the threshold for distortion than they are, and often that comes from the squash of compression. It's so hard to undo that. So trust that I can get it as loud as it needs to be. We need to have some headroom to do our thing.

Q: I assume then you're comfortable going back to the mix engineer and saying, "I need this, this, and this"?

Yes, assuming that the budget hasn't been used up. Often, they can accomplish something better in the mix phase. There are some mix guys who will do some mixes and come for a quick visit to listen in the mastering studio to get a sense of how it might turn out, then go back and adjust their mixing. I saw this happen periodically at Sony Studios when I worked there. They had a big facility, where they had mix rooms on one floor, mastering rooms on the other, so it was easy to pop back and forth. I remember one instance for a particular jazz record, they were actually mixing and piping the mix into the mastering room, and then going and listening to it there in real time so they knew exactly how they needed to adjust their mix. Most people do not ever have that luxury. It was great for that particular client. It wasn't my client, it was somebody else's client, so I won't name names, but they were really careful about doing that, and they had the means to be able to do that.

If you look at a lot of new records that come out now, they have been mixed by four or five different people. In the "old days" you got something that was recorded all in one place, maybe overdubbed in a couple places, mixed all in one place by one person. So you've got somewhat of a coherent product to work with. It was a more cohesive package and it was one guy who mixed the whole thing, so it was easier to go back to that person if need be. Now, you can't always go back to the mix engineer, because it's one of four or five people, and they've moved on by time you get it. That is a challenge for mastering engineers that we didn't used to have. Now, sometimes you can get 10 tracks mixed by 10 different people. You have a whole new challenge of not just how to make each track sound as best it can, but now you have to make them sound like they're related, and very often they're far from related.

Q: Does that come through equalization or anything in particular?

No, because it could be all over the map depending on what the issues are. So, that really does push the envelope of your skillset. I think in that case, it's another place where it really comes in handy to have some under-standing of the process of how these tracks come to you. Does something sound a certain way because it's this particular engineer's style? Is he/she known for using a particular piece of gear? Or is it the particular room? Or because this track is a little bit more "hip-hop" than this other track? Or more dynamic? It could need EQ, it could be ambience, it could be compression. Maybe it's harsh converters you are fighting. I think the bag of tricks that you have to have is now a bit wider ranged. Compression, ambience, dynamics, signal chain. It all plays a part.

Q: In terms of formats, are people releasing fewer albums now, or is that overstated?

Not overstated. I get asked to do singles, or Eps, probably way more fre-quently than I used to. The formats things came in on used to be ½-inch 30IPS, maybe it was Dolby, maybe it wasn't Dolby. Now you're getting a wavefile or some other digital file format, or sometimes they're send-ing you the Pro Tools session itself. They could bring in a hard drive, or it could be on a USB stick. The other big thing that has changed a lot is that they're not always coming to you [in person]. They're sending some-thing to you. The whole personal nature of it has been somewhat removed; it's a lot more remote communication, which can be good and bad, quite frankly. It can be nice to not have people over your shoulder commenting. It's like, "Leave me alone. Let me do my thing," then you can comment. So, there is that aspect, and I don't mean that to sound as cold as it is, but sometimes you just need to do your work, and then you can ask questions later. But it makes it a lonelier process, which makes me sad. I also get requests from other countries more now, so the language translation can make getting to an understanding of direction more difficult.

Q: If you would be willing to do so, could you speak about gender, and gender in the industry at all? We have many students who are young women, and they want to enter the field. They communicate that they feel that there is an added difficulty, one, in being taken seriously and two, even just in getting the gig. Do you have any advice?

When I started out I went to engineering school and there were two women out of 20 in my audio class. Now there's probably half a dozen, maybe there are eight in a class. I absolutely think it's changing, but I'm not going to pretend that there's no difference. I have seen, over the years, that it gets a little bit easier. But I won't pretend that it's okay. For example, I can't tell you how many male engineers that I worked with were dopes, didn't have their act together and didn't know how to do one thing or another, and they kept getting gig after gig. That doesn't fly with women. As a woman, you have to know what you're doing, you just have to. You can't fake it. You can't bullshit your way through like guys can, because nobody's having it.

Q: Do you feel like you're being scrutinized?

Yes. Even the most enlightened males, no offence, are assessing you in a way initially that a male counterpart would not be assessed. I have found, not to be totally discouraging, that a fair portion of people, once you pass that assessment, become okay with it. There are still some who don't, and there have been situations (and I think that this could be a whole other breakout discussion) over the years where I'm certain I was taken off a session or I didn't get asked to do a project because someone involved didn't like the idea of a woman engineer. I absolutely think there are times where being a woman did negatively affect me. The best advice I could give to any young woman is to make sure you know your stuff and don't pretend to know something that you don't, because it'll trip you up. They won't forgive you, where they would forgive a guy for pretending to know something when they really didn't, but they won't forgive you. I think you've also got to find a fine line between being confident and assertive, but not being perceived as a bitch. And that is not easy. I think what I'm saying here is not exclusive to the music business, either. I think women in any male-dominated field will give you the same sorts of answers. You would think that the music business was all hip and everyone is much more enlightened, but there's still some of it out there. Even after all my years, I still encounter it. But I will say it is improving.

I would also say try and be a sponge. Any older or more experienced person you come in contact with, try and engage them and find out what they know. Part of what is sad about the demise of the recording studio, and this applies to everybody, but I think it hurts women even more, is in the heyday of studios, you had a community. You went and were a gopher, wherever it was, in some big studio. A whole host of different kinds of engineers and artists came through there. You were able to see a lot of things and learn a lot of things; young engineers today don't have that. You don't know what somebody else is doing because they are in their basement, that's where half of it is happening. Who do you learn stuff from if you can't get to them? How do you find a quick and easy way to do this or that thing if you can't walk down the hall and check out what's going on in the studio next door? So it's hard, and it makes it hard to learn and get better at your skills. Not to mention, how do you make contacts, market yourself and get gigs? Just to develop a diverse skillset is difficult. That's bad for everybody, but I think it's especially bad for women because I feel like we have to be so much more infallible.

Q: We talked to one guy who had a 45-minute philosophy on mid/side. What's your philosophy on mid/side?

I don't have a philosophy on mid/side. Some engineers can be intense and completely, totally, transfixed on solely mastering. That's not who I am. That's not how I've come to this place. Like you've said, my career path has been very non-traditional. Sometimes I even feel uncomfortable calling myself a mastering engineer because I feel like those kind of guys and gals are mastering "gurus," and they would likely not consider me one of their peers. You know what you know and have your range of experience, and you use those skills for the task at hand. In the end you hope you come out with a great-sounding product.

Randy Merrill

Randy Merrill is a senior mastering engineer at Sterling Sound in New York City. He works on recordings of a wide variety of musical styles, including pop, rock, electronic, dance, hip-hop, jazz, Latin, country, and classical. He works with clients from major and independent record labels as well as independent artists from all over the world.

Randy first developed an interest in music technology as a student at Jamestown Community College, where he earned an associate's degree with an emphasis on music. He earned a bachelor's degree in sound recording technology from SUNY College at Fredonia, where he studied music, audio engineering, and acoustics. After graduating from Fredonia, he worked at the Eastman School of Music in their recording arts department, where he honed his skills in live and studio recording and mixing and live sound reinforcement.

In 1999, Randy moved to New York City and worked as a freelance engineer, assistant, and technician. He spent five years as a technical engineer at Avatar Studios (formerly known as Power Station). During this time at Avatar, he was involved in the construction of their mastering facility, which ignited his desire to pursue his own career in mastering.

In early 2006, he joined Scott Hull Mastering as an assistant and technical engineer. Shortly thereafter he started attracting his own clients. When Scott purchased Masterdisk in 2008, he became a staff engineer there.

In 2013, Randy joined Sterling Sound as Tom Coyne's assistant. In 2017, Randy won Grammys for Album of the Year for Adele's *25* and Record of the Year for Adele's *Hello*. In addition, he was nominated for Album of the Year for his work on Justin Bieber's *Purpose*.

Randy lives in the New York City metropolitan area with his wife, Aya. Aside from mastering, Randy enjoys traveling, spending time at the beach, and playing music at his church.

Q: What is mastering to you?

Mastering for me is all about giving a recording that final polish before it gets released. A big goal in mastering is to make sure that the sound translates well onto multiple listening systems. I aim for a sound that is going to feel good in the car, on the radio, in headphones, on iPhones and other portable devices, CD players, and home stereos. Whatever you're playing it back on, I want the music to sound well represented and engaging. Every environment will sound different, and as an engineer I don't have control over the listener's environment, but at least I can aim to give an overall balance to the sound in hopes that it will translate reasonably well. Whether it's adjusting the bass, treble, or midrange EQ, or controlling the dynamics, I try to get it to feel relatively even, yet exciting and emotional, bringing the most out of the music. This applies to mastering single songs and albums. The simplest goal in mastering an album is to make it so the listener can play through the whole thing without having to adjust their playback settings, such as their volume or EQ. They should be able to listen through the whole album without anything unnecessary jumping out, or feeling strange or out of place, so there are no sonic distractions, unless it's intended. Aside from this, mastering should create a sound which enhances the character of the music, making it more engaging, emotional, and exciting. That, in my opinion, is what mastering is.

Q: What made you get into mastering in the first place?

For me, it was a process of elimination. I came out of a music school that had a recording program. I wanted to be a recording engineer and music producer. I was very much into playing different instruments, and I just loved being in the studio. When I moved to New York City in the late '90s, I found myself at a point where the music and studio businesses were drastically changing. A lot more studio work was DAW based. It became more about editing and fixing, and to me what seemed like "word processing" for audio. I also figured out that my personality didn't work well for being in the same room with the same people for 14 hours a day. Between these two things, I realized, "I don't think I can be a recording engineer." I love recording, I love putting up mics, and getting sounds, and working on music in a creative sense, very much. I still do a little bit of that these days, but I knew that I couldn't make it into a career. So, at that point, I thought, "How can I use what I know, to make a living?" I had been in New York for less than a year, and I was 25 years old. Fortunately, I was able to get a gig as a house technician at Avatar Studios, which at the time was one of several multi-room recording studios in New York. It was The Power Station from 1977 to 1996 and was designed and built by Tony Bongiovi. A lot of big rock and pop hits were recorded and mixed there. In 1996, it changed ownership and took the name Avatar Studios. They have four rooms with large analog consoles, plus two smaller editing and mixing suites. It is now one of the very few large format studios remaining in NYC, and it was recently acquired by Berklee College of Music.

I got a gig there as an entry-level technician working in the maintenance department. I was reasonably handy, but I wasn't an electronics engineer by any stretch. I basically learned how to do it on the job. I'm not a gear designer, but most of the time if I find something broken, I can figure out what's wrong with it, and fix it. I learned the hard way, often by breaking things more in the process, before eventually figuring out how to fix them.

In 2004, we put together a mastering suite at Avatar for a gentleman named Fred Kevorkian, who was going independent at that point. He had his own gear and was looking for a room to rent. We ended up doing a makeover on one of our spaces, and it became his home. As we were putting it together, I was really fascinated by all of it. He was gracious with me and let me hang out while he was working, and I thought, "I really like this, maybe I can do this!" I had never thought about mastering as a possibility because I thought you had to be gifted with special hearing to be a mastering engineer, but it's like any specialty in that it's something you have to learn and practice in order to do it well. With this desire in mind, I continued working at Avatar for about a year while looking around for opportunities to pursue mastering. I met Scott Hull in late 2005 at the time he was going independent and putting together his own studio. He had worked at Masterdisk, Classic Sound, and at The Hit Factory NYC until it closed. I worked with him for a few weeks while he was getting started in late 2005, and he hired me in early 2006 as his full-time assistant.

Because I had the maintenance and recording backgrounds, I was able to fill a lot of gaps in what he needed in an assistant. I became his engineering assistant, production assistant, maintenance technician, and somewhat of a facilities manager. Just about anything that needed to get done in terms of the infrastructure of running a studio, I had to figure out how to do it. I had to learn how to get a PBX phone system to work, so that we could have a receptionist, a studio phone, a lounge phone—stuff like that. Of course, at that point I was learning mastering from him, and about a year after I started there, I began having some of my own clients and doing sessions on my own, and it's just grown from there. I moved with him when he bought Masterdisk in 2008. My duties there continued in the same manner, both in terms of being an engineering assistant and taking care of some of the facility details, alongside working with my own mastering clients.

In early 2011, I started focusing exclusively on mastering, and that was how I continued at Masterdisk through the end of my time there. I moved to Sterling Sound in late 2013 when an opportunity opened up to be one of Tom Coyne's assistants. At that point, my wife had been working with Tom for about 10 years as his assistant, and Tom had gotten so busy that they needed another one. When I started there, I spent half of my day working for him and the other half working for my own clients. It was amazing to see "The Sterling Level" of how things happen. The sheer volume of work that goes through is just mind-blowing. Sterling is a true international company, working on hit recordings that come from around the world.

Q: You say you're still learning. You're still on the job, really, aren't you?

Every project that comes in teaches you different things about what you have to do in order to get it sounding just right. The most important thing to me is keeping an open mind and not going into any project with assumptions. To me, one of the biggest responsibilities I have as a mastering engineer is to have the perspective of, "Do I need to do next to nothing, and preserve what this is because it's so good, and I don't want to do anything to distract from it?" Or "Do I need to completely reinvent it because it's so far off?" Or is it somewhere in between? I have to make this decision all day, every day. If I work on 10 projects in a day, they can fall all across this spectrum.

Q: How do you deal with clients and telling them things about their music? Let's say you have to reinvent the material, as you were saying. How do you start that discussion?

Honestly, I assume that when clients are sending me their recordings, they're giving me the best mix they can do with the tools and experience they have, and they're hiring me for my opinion on how it should ultimately sound when it's finished. So no matter what I have to do, whether it's a little or a lot, I have to stand by my opinion of where it needs to go. At first, this didn't come easily to me, but it was something I learned during my time working with Tom Coyne. He was just fearless. He, and now I, have a certain grace with it, too. If the client doesn't like it, I'll redo it, of course, as did he. There has to be some amount of pushing boundaries just to see where the music can go in order to give it that "wow" factor. As far as the conversation goes, sometimes there isn't much conversation, because it's hard to tell someone that their mix is really far off. Just like mastering, it takes years of experience to learn how to mix well. So how do I explain to someone, in a single conversation, what they need to do to fix the problems I hear in their mix? At times I have done so, and that conversation has gone well, and other times it hasn't. I've had times when

I've talked with the clients about making changes, and the new mix I get isn't better, it's just different, and still requires a lot of work to get it to the finish line. In many cases, I feel like mastering has become "mixing part two" in addition to mastering, so I see it as my job to do whatever is necessary to get it to be the best that it can be. What I do is ultimately a customer service job, so I'm always willing to talk with clients about what they want to accomplish in mastering.

Q: Let's say you get a job in, are there any key processes you go through? How do you approach each and every job? Is there a routine you go through?

For an album, I'll go through and sequence the songs first, putting them in order and doing the fades and spacing. I work in Sequoia, and I keep everything at the same sample rate as the source. If it comes in at 88.2 kHz, then I EQ it at 88.2 k. If I go analog, I record it back to the same session at the same sample rate. If I stay digital, it's already in my DAW at the source sample rate. I work at the highest resolution of the mixes in order to be prepared for all of the possible delivery requirements. I do it this way rather than having to do a separate mastering pass at 44.1 kHz for CD, and then a whole other pass at 88.2 k or 96 k, or whatever, for the higher resolution delivery formats. I have found that it sounds better doing it this way, after doing some blind testing between this method and re-recording it at higher sample rates. To me it sounds better if I keep it at the higher resolution until the very last stage, and then "down-sample" it for whatever I need to deliver.

For an album, once the songs are in order, I'll do a basic levelling on it to get the songs feeling like they're around the same volume as I go from one to another. I do this mostly by focusing on the midrange. If it's a vocally driven recording, I'll try to make the vocal feel equally loud across the whole album. I'll use that as a basis for figuring out approximately where the song should sit level-wise, in terms of its relation to the songs before and after it. I'll get an overall view of the album that way. Then I'll dig in and see what kind of processing is going to be necessary. A lot of it is about making mental notes, like, "This song has too much low end. This one doesn't have enough. This one needs some compression, this one doesn't. This one is a little dull, but this other song is plenty bright, so I've got to be careful with that one." Then I'll figure out what the chain is going to be. I'll try to see if there's any one thing that works for a lot of it, such as a plugin limiter, an EQ, or a converter if I'm going analog. I've got two D-A converters I can choose from on the playback side. They're different flavours. They're just EQs, really. Everything is an EQ, whether we realize it or not. Just going through a piece of gear or a plugin can change the EQ or colour the sound. I figure out what's not going to hurt the recording and what's rather going to enhance it in a musical way. After I've figured this out, I'll go through song by song, making further tweaks to get it to where I think it should go. Once the decisions have been made, if I'm going analog, I'll record it in, or if I'm staying digital, I'll capture it. I'm always referencing back to songs I've already done to make sure I'm in line with where my original thought process was for the album.

I try to work fast, because to me, mastering is all about the first impression, that first thought I had when I heard the song for the first time. Early in the process, I'll develop a mental idea or a mental picture of what I think it should sound like, and then I just spend the time taking it there. I try to work fast so I don't lose that first impression. If I get deep into something and get lost, I force myself to move on, even if it means I have to come back and tweak it later. To me it's better to move on rather than to get burnt out on it.

Q: You referred there to analog and digital. You don't have any preference particularly, or it's just whatever works? Some engineers we've been interviewing will spend the time doing an analog session, because they know they're getting paid at a proper rate. If they've got e-mastering stuff coming in over the Internet, which I'd imagine for most is everything these days, but it makes them say, "We'll just do digital through and through, we have a chain. We'll do x, y, z. . . ." Do you have any sort of delineation between the two?

Absolutely not. I do it based on what the project needs, because that's why I'm being hired, to give them the best possible sound. If it means staying digital, I stay digital. If it means going analog, I go analog. Sometimes it's different within the same project. I might do 80% of an album digital, but two songs sound better going analog, or vice versa. I'm not afraid of mixing and matching it on the same album based on what I think it needs. I definitely don't do it on cost whatsoever.

Q: Do you have any quirks in your practices? Anything unique you would say you do that's innovative, or something that's unique?

I'm not sure if this is innovative or unique, but after I have gone through an album listening on the speakers, and I think I have it sitting the way I want, I will spot check it on headphones, and again on the speakers at a low volume, and will make adjustments if I hear anything I think needs to be changed. I sometimes listen in mono and do the same thing, mainly to make sure the vocal sounds as good as it can.

Q: Me [the interviewer], for example, I tend not to compress things; I limit things.

I totally get that. I can agree with that. The amount of times I actually compress something is a low percentage. I'm definitely more into limiting, EQ, and volume automation to get things sounding full and even. Sometimes I think, "Okay, this needs some compression" but then, after I A/B with and without it, I often find it to have more life and excitement without it, so I end up not using it. Sometimes I will use multi-band compression just on the low end if I need to get a kick or bass instrument under control, when cutting bass frequencies isn't enough. I just do whatever I have to in order to make it sound the way I want. I'm not afraid of throwing the kitchen sink at it if I need to. I find that much of achieving loudness is about EQ balance. A recording that is balanced in terms of EQ is going to sound loud naturally, because it has a fuller frequency spectrum. If you use limiting in a tasteful, creative way, you can get away with a fair amount of it, and the music stays intact without changing the mix balances too much.

Q: Do you do much parallel limiting or compression?

It's not that often, but sometimes I use parallel compression, when the mix is overcompressed, or when the transients are buried and I want to restore some life or punch to it. I don't do any parallel limiting though. Maybe that's the next thing I'll try!

Q: Are there any go-to pieces of gear you tend to say, "I always have that in my chain?"

I don't have anything that's in all the time. In terms of plugins, I use the FabFilter Pro-L often. That was a big change when it came out on the market, in my opinion. To me it made a huge difference, a big leap forward in transparent limiting and getting cleaner sounding volume at louder levels. I really enjoy using it. I've got several other plugins I use for limiting, but the Pro-L is a staple. I try to keep an open mind about what I use and go into every project with a blank slate. I don't go in thinking, "Today I'm going to use X, Y, and Z, because today is 'X, Y, and Z Day.'"

Q: What are your thoughts on "The loudness wars," generally speaking?

I think certain kinds of music can sound great loud. I think that there can be a certain visceral quality to it, when it's just absolutely smashed and pushed to the limit, because it causes an emotional reaction that you can't get any other way. However, there is plenty of music being made that doesn't need to go there, and there's no benefit to pushing it that much. On certain kinds of acoustic music that I work on, like jazz or folk recordings, where the clients are asking it to be loud, I think, "No, it doesn't need to be loud. You've got to let this breathe." To me the music has to dictate what it is, where it can go, and how loud it should be. I'm not going to be somebody that says that there isn't a place for it, because I think that there can be a place for it, and it is appropriate for certain kinds of music. It is hard work to get something loud and still have it sound musical. I've heard great sounding loud records. And the engineers who have mastered them have spent the time trying every tool under the sun, from analog to digital, to figure out how to do loudness well and make it sound like music. To me, that's an approach that I respect more than those saying, "oh, loudness is just ruining music."

Q: Along those lines, what do you think the future of high-definition audio is for the consumer? You've mentioned that if something comes

in at 88.2 k, you'll do it at that, but how are you delivering to "Mastered for iTunes?" Are there lots of expectations from clients for that kind of stuff?

Yes. For "Mastered for iTunes," we're delivering 24-bit audio. If the project comes in at 96 kHz, we deliver it at 96 k/24 bit for iTunes unless otherwise requested, and we'll drop the level the necessary amount for a good conversion on Apple's end. I suspect (and hope) the reason why Apple is asking for "higher-res" files is so that they will be able to deliver uncompressed, full resolution audio in the future. The restrictions to date have been internet bandwidth and making downloads more efficient, as well as the limitations of portable music players, like iPhones. Apple seems to care a lot about battery life and storage space. For them, an AAC file is much more conducive to their devices in terms of getting music downloaded, stored, and played back on them. Hopefully, that will change, so that we will be able to easily download 96k files onto our portable devices and be able to listen to full resolution audio wherever we want. I'm hopeful about this for the future. Is high resolution audio necessary? There's definitely a lot of debate whether sampling rates above 44.1 kHz are necessary. It's true that human hearing only goes up to a certain point in terms of frequency response, but I think that there are details which can be enhanced through sample rates higher than 44.1 kHz. So, to me, it's advantageous if we move towards that for the purpose of better enjoying recorded music. However, it often ends up being more about convenience for the consumer, what they're going to want, and what they're willing to pay, ultimately. That's a challenge in and of itself, getting people to pay for recorded music.

Q: What are your current favourite monitors?

At Sterling I use a custom-built system designed by Ted Jensen in the '90s for Tom Coyne. It is a four-way active system that has Dynaudio tweeters, ATC midrange drivers, and a dual pair of Scanspeak mid/bass drivers, all powered by Bryston amplifiers. It also has two Rhythmik subwoofers to extend the low frequency response. At home, I have a pair of Tyler Acoustics D112 loudspeakers. Tyler is the builder, and he sells directly to the customer. He's based out of Kentucky, here in the United States. He builds different models of hi-fi speakers which several independent mastering folks have embraced. I bought a pair because I heard a lot of good things about his designs. I enjoy them in my home listening setup and would feel comfortable mastering on them if I had to.

Q: Looking at the Tyler Acoustics stuff, and I've noticed in the pictures that they look quite big.

In terms of size, the D112s are in the middle of his line, but he has all different models. He has some which use an augmented MTM design, with a tweeter in the centre, and mid bass and bass drivers above and below the tweeter, with a recessed baffle to time align the tweeter. He also has flat baffle three-way speakers, like the ones I have, which contain a 12 inch bass driver, and a 5 inch midrange driver, and a three-quarter inch dome tweeter. I use the Sterling setup for work and the D112s for listening at home.

Q: What do you think the future of mastering is? You've mentioned that Sterling is really busy. What do you have in store for us?

I'm hopeful that mastering will continue to be a viable and profitable business. I think more than ever it's necessary, because there's so much more content being created now, and the creative tools are so much more available to people. Anyone can buy a laptop and Pro Tools, and a USB interface, and a microphone, and make a record. I think more content is being produced than ever before, but more of it is being produced in less-than-ideal environments. To me, I feel like mastering will continue to be necessary because of that, in order to bring a real objective viewpoint to a project before it gets released. There's speculation that someday a plugin or an automated service will replace the mastering engineer, but for now I think the mastering service (by a human being) is crucial.

Q: There is something out there, but we're probably best to not mention it in your interview. That's something we are going to have to discuss at some point in time.

I value the diversity of people involved in a project. For me, if I was an artist and was producing a record, I would want to go to a studio that I know has a good reputation, work with an engineer that has a good reputation, work with musicians that have good reputations, and work with a mastering engineer with a good reputation as well. It's valuable to have all of these different viewpoints involved in the process—I think it only adds value to the project because everyone has their specialty and valuable input. When a person who is the artist and is determined to be the producer, and they're the mixing engineer, and they're the mastering engineer, I think there are limitations to what they can do because they're trying to wear all of the hats. I think the recording comes out better when it's a team effort.

Bob Katz

Bob Katz (Robert Katz) is a three-time Grammy-winning mastering engineer who has been an AES member since 1975. Fluent in English, Spanish, and French, he's played the B♭ clarinet since age 10. He is the author of numerous articles and equipment reviews published in *Byte*, *dB* magazine, *RE/P*, *Mix*, *Resolution*, *AudioMedia*, *PAR*, *Stereophile*, the *AES Journal* and at the digido.com website, which hosts a comprehensive Audio FAQ.

Q: What you do is musical. It is an aesthetic craft to you?

It is often an aesthetic craft, while it was not always regarded as such. For your readers who have not used a good mastering engineer, I cannot over-stress the importance of letting them know that mastering can be an amazing experience. It's having a second listen to their work by a consummate, experienced professional, [and] they really may need to have it done. This reminds me of an automated online service which is being misrepresented very strongly by commercial interests. They even received 10 million dollars in endorsement from TuneCore. What did George Bush say? "Fool me once, but I won't get fooled again."

Q: In your mastering, are there any things in particular that you find yourself listening for? For instance, is there a listening technique?

The answer is yes and no, because I need to recognize that each recording and performance is unique in order to do my job. I shouldn't come in with a prejudice that says, for example, the vocal always has to be at this level. On the other hand, there are many genres where the vocal is often required to be at a particular level, and I should be aware of those genres. If I hear an exception, I should be able to say, "For the art of this piece of music, should we obey the genre? Or should we go with what that piece of music seems to be trying to tell us, that's unique and different from the genre?" So, in summary, we have to be open-minded. At the same time, in terms of our creativity, I think we all have to admit that we either consciously or subconsciously have an archetype in our heads of something that sounds, or may sound, or have aspects of sound, like what we are working on today. We may notice that this particular piece of music doesn't quite have the qualities of the archetype, and we may find it to sound better with some of the attributes of the archetype. Only through experience can we recognize that need. For example, if a piece of music has a decent rhythmic bass and we like that, through experience we might decide, "Well, gee, this music would be even more danceable with more bass." Assuming that "dance-ability" is what we're going for. Only through experience can we know that more bass will work for this recording, and hopefully we're right better than nine times out of ten. Sometimes the band will come back and say, "Hey Bob, there's too much bass," and I was wrong. That can happen too. Only through lots of good experience can we get 90% or better in terms of what we choose to help a piece of music.

Q: We wonder if you might think it's fair, then, to say that there's almost an element of curation to mastering.

I would have to say yes, but at the same time there is resistance to this concept from people who say that "Mastering engineers are being too creative, and they should let off. They should just try to fix some minor EQ problems, if there are any, and let it go." There are recordings that I get where curation is exactly what I do. An increasing number of recordings that come to me need that curation; I have to be able to know the difference. I have to be able to say, "Well, this is an incredibly creative and wonderful piece of music. Should I leave it alone? Is there anything that I can do that would take it further in the direction that they're trying to go?" All these are good thoughts, but notice that they're generalizations, because for each project it is different.

Q: We're interested in your artistry and experience. Could you tell us more about this approach?

Okay, good. Then let's talk about something that I call. . . . Are any of you a fan of Star Trek? You know what Captain Kirk's favourite saying was? "Bass—the final frontier." It is the final frontier, because it seems bass is the most difficult thing for recording and mixing engineers to get right—but it doesn't have to be that way. There are plenty of good monitors that tell you what the bottom end is or should be doing. There are plenty of ways today to get a monitoring room in good shape at reasonable cost, including digital corrections and therefore to some degree the room response. On the one hand, I'm kind of glad that these clients have problems, because then they can come to me. On the other hand, I'm a big fan of pointing out that the quality of your monitoring is directly proportional to the quality of your work. Keep that in mind when you talk about "Bass, the final frontier."

You don't necessarily have to wait—and you shouldn't wait—until the mastering engineer gets a hold of it. For example, if there's a problem with the bass instrument, the bass drum may have unfortunately an opposite problem, one that can't be cured in mastering by the same equalization that fixes the bass. There are other tools besides equalization. This is why the first thing I recommend is to come to the mastering engineer as early as possible and send him or her an example of your mix in the making. When I get an example mix, I can advise the client as to whether it's ready for mastering or whether there are some things that they can do. At that stage, I try to judge whether they're going to be able to fix it or whether it's better for me to fix it with my tools. I may suggest, for example, three or four stems: vocals, instruments minus bass, bass, and sometimes bass drum. Stems can be a tremendous help, because, for example, I can apply an EQ that can clarify the bass instrument without getting in the way of the other instruments. This is often not possible when I get a full mix. This seems to cross the border between mixing and mastering, but it's more a case of mastering to bring out the artist's intent. This situation is happening more and more these days, as there are more inexperienced music producers. The number of times that I recommend stems is most often when there are problems with the bass instrument, reflecting the inferior monitoring used by so many producers today.

I might go to an extreme in repairing a bass, with the approval of the mix engineer, of course. There's a hard rock group from Australia called Icecocoon. The leader of that group, Owen Gillett, is a very experienced mixing engineer. He's a graduate of and also a professor at SAE. It's not as though he doesn't have the abilities to mix, it's just that the challenge that was presented with his group is an über challenge, one that would challenge anyone except the most experienced and creative producer and mastering engineer with the finest monitoring. The reason is the bass instrument in this hard rock group is what we call "drop tuned." The low E was tuned to low D, and in some cases low C. This can produce a kind of "rumbly" sound which has no definition. I was able to not only meet that challenge but to transform a recording where the bass was originally literally inaudible to a recording where you can hear all the notes playing properly, coming through with punch and clarity. I also was able to help the bass work well in tandem with the bass drum. This approach just enhanced the whole impact and strength of Icecocoon.

This begins to cross the border between mixing and mastering. There are mastering engineers who have never mixed in their life. They wouldn't dare to get stems. The question, of course, is "Are stems really mixing?" If it's just a case of vocal versus instrumental, hopefully the mastering engineer can handle that. It's true that there are people in your audience, if they're budding mastering engineers, who would say, "I've never mixed." The advice I would have to give is, "Maybe you should try mixing for a while before you take on stems." Because not every mastering engineer is prepared for that, and also you may have to put on your mixing hat. The question that comes up is, "Should you have a separate day to work on the mix of those stems, and separately from that you put on your mastering hat?" There are definitely two different hats that you have to wear. Maybe some people are schizophrenic and can do it, but in a session as complex as Icecocoon, I have to have a separate mixing session, think about whether the balance between instruments is working, then switch over to mastering. In that case I spent a few hours wearing the mixing hat, then took a break and returned to do the mastering session. I can't wear both hats at once. Maybe there are people who can do that, but I can't.

Q: Would it be possible to ask you to elaborate a little bit on any specifics that jump up about why that is?

I think it's not only that you can't do it, but also that you shouldn't do it. The goals of translating a recording to the listener for mastering purposes are different than the goals that the mixing engineer works on to make a good mix recording. When you're mixing, it would be nice to concentrate on the internal dynamics of the recording, but that is one of the last and most difficult parts of the mix to convey. You have to be working on getting the internal balance of the drums the way you want them to be, you decide whether you want the snare to be standing out or whether this is not a snare tune, and so on. That's not my job as a mastering engineer. My job is to take and polish, if necessary, what you did and to respect all the philosophy and approach that you gave when you mixed the tune. The mixing engineer shouldn't be concerned if he trusts and knows that I have the reputation to accomplish what I've done while preserving it, and yet possibly enhancing the philosophies that you produced in your recording.

Let's get back to Icecocoon and talk about what I did. The first thing is that I have an incredibly accurate monitoring system. It's also very musical. It doesn't sound so analytical that it's fatiguing in any way; it's just really accurate, and I work really hard to make sure that it is accurate. With that accuracy, I can hear what the problems were, in this case in the bass instrument, and especially if he's going down to low D or low C, how to translate that to as many listeners as possible. However, we might ask, "Well, gee, if Bob's system goes down to the centre of the earth, does that mean it's going to playback on a MacBook Pro?" The answer is, it translates as well as possible to the MacBook Pro, but there are things that might be missing. It will translate even better to the MacBook Pro than it did before I got hold of it. Mastering has translated very well to any reasonable medium or small system. Technique-wise, if I recall, the bass was recorded direct. I usually find that in hard rock a direct injection box cannot work that well. I believe that if you want a bass to speak properly within a rock band, you should think about what it sounds like in a live performance when it comes from the amplifier. In my case, and this is very radical, I transformed the bass to bring out its harmonics and note defini-tion: I used an amp simulator. After 48 years in the business I guess I can dare to be radical. So I used the amp simulator, and that blew the mind of the producer; it really did. He's counting on me, and he's coming to the mastering engineer to help him take this difficult mix to its conclusion. People come to me for what it is that I do well, and they trust me. Knock on wood here, most of the time they come out satisfied. I have my failures, but most of the time they come out really happy.

Q: You mentioned earlier the accuracy of your monitoring chain. You've worked very hard to create a monitoring situation that is trustworthy and accurate. We're wondering if you're willing to take us through what your monitoring chain is, why you chose the pieces and what they contribute.

My monitoring is unique. Very few mastering engineers employ response correction in the way that I do. First of all, I have a reasonably good room to start with. It could be better, but in some aspects it's extraordinary. The room is well treated. It's not too live and it's not too dead. The walls are treated with staggered pieces of hard Sonex, so one wall will have a Sonex piece while on the opposite wall there is a hard surface, which is sufficient to get rid of any flutter echo without overdamping the room. The room goes up to 23 feet high in the back! It has a cathedral ceiling, which com-pletely eliminates floor to ceiling reflections and resonance issues. There's trapping to help deal with bottom-end issues, but it's not overdamped. I use both passive traps and active traps from PSI audio called the "AVAA." I've looked at the Schroeder curves and made them to be extraordinarily even, octave-to-octave-wise.

The listening position is a reflection-free zone. If you read Floyd Toole's book, he doesn't believe in the necessity of reflection-free zones, and with all due respect to an extraordinary scientist, Floyd Toole, whose work I greatly respect, I believe that the reflection-free zone helps me to get better work. Measurement-wise, there are no specular reflections greater than 15 dB below the initial impulse for the entire period of measurement. What that means is that the direct sound from the loudspeakers is not interfered by the early reflections.

The main stereo speakers are Dynaudio Essence M5P, which are 8 feet tall, weigh 300 pounds and go down to 27 Hz. Those are supplemented by my custom crossover with JL Audio Fathom F112 subs, which extend the response down to 15 Hz. For stereo, I'm using four channels of DAC. Two DAC channels feed the stereo subs, and two DAC channels feed the mains. For surround, the total number of channels for 5.1 is eight. So I have eight channels of DAC. The left surround is crossed over into the left subwoofer, and so on, preserving the sense of ambience and depth for all the channels. Although we cannot detect direction below 100 Hz, we can sense ambience and spatiality.

All of this is corrected with a system that I believe to be the most transparent digital room correction system on the planet. I spent over a year comparing the sound of this digital correction to my previous basic analog crossover and decided that the digital correction is the clear winner. That system is by a company called AudioVero. They make two products, one of which is an extraordinary analyzer that psychoacoustically measures the response of the speakers. For reproduction, a piece of software called Acourate Convolver that convolves a set of filters. Acourate is FIR and so can correct phase and impulse response as well as the frequency response. It improves the impact and transient response of even this system, which was no slouch to begin with, because it does not overcorrect. There is no limit to the number of filters. Unlike most correction systems, there is no compromise, all the anomalies are corrected, and Acourate can distinguish between room problems that should be left alone and loudspeaker problems that should be corrected, transparently and invisibly. But this is a rare and radical solution for mastering, and that's why I spent over a year doing A/B comparisons. Acourate Convolver is a 64 bit floating point system running in a dedicated separate computer that runs my monitor system.

Some might object and complain, "Gee, we shouldn't touch that. Why are you putting so many filters and equalizers in your system? That's totally crazy!" But, wait a minute, loudspeakers themselves are compromises. They already contain equalizers, resistors, inductors, and capacitors to optimize the frequency response. What I am doing is an extension of the loudspeaker design, if I do it well. I also perform a linear phase crossover between the subs and the mains, which you can't get in the analog domain. There are no dips or peaks in the response in the crossover region.

The only downside to Acourate is the latency of FIR filtering. So when I edit PQ code or do video sweetening, I switch to a zero-latency alternative that still sounds good but pales in comparison to Acourate.

Q: Can we talk a little bit about an analog processing chain you may use? We also wanted to get your opinion on conversion, how you yourself evaluate conversion.

There are mastering engineers who use coloured converters on purpose, which is not my philosophy. I would rather let whatever analog processing that I choose to use provide the colour. Even the word "colour" is controversial, because let's just say that it's shorthand for the compressor, expander, equalizer, or another processor that you might apply.

So I believe in having accurate converters. I've gone through a bunch of converters, and, obviously, in mastering, 99% of our sources are digital these days. There was a period of time where mastering engineers getting analog tapes would run it though their chain on the way to an A/D and call it a done day. Even when I was getting a lot of analog tapes I might do some initial EQ in the analog domain, but then I might do further work later in the digital domain.

There are many models of good converters. I have at this time settled on the Prism converters because they're the closest to transparent. How do you know if a converter is transparent? You take your source on your DAW and you monitor it directly through your monitor converter, and then you insert in the middle of that chain a D/A going into an A/D. The winner is the insert that sounds as invisible as possible, where you can't tell whether it's in or out. There is no transparent converter, but the Prism comes closest for me.

For processing, now I have and was the second person in the Americas to discover it, an analog EQ from a company with the weirdest name in the world, called "Bettermaker," the model EQ 232P. Digitally controlled analog is very controversial, but I find the Bettermaker to sound very pure. It does Pultec curves better than Pultec! I also have two pieces by Maselec, the MLA-4, which implements downward compression or upward expansion in three bands, a unique analog piece, thank you, Leif Mases. And I have his MEA-2 Equalizer. This is supplemented by a third analog EQ, the Dangerous BAX. I also have outboard digital, the TC Electronic System 6000, which has many algorithms that I use and works in surround. All supplemented by many available plugins, including my own invention, the K-stereo processor, available from UAD.

Q: Are you still using your Weiss equipment?

I have sold both my Weiss DS-1s; the hardware has now been superseded by plugins. I contributed greatly to the performance of the Weiss DS1 Mk2 and Mk3, and now they've put out a Weiss plugin by Softube that is the same algorithm. I also use a plugin called "Essence" by DMG which overlaps the Weiss functionality and puts its own spin on it.

Q: Many engineers we've spoken to who were initially interested in multiband compression have come to give up on it as a regular activity. We'd be really interested to hear your thoughts on this.

You're right: Multiband can be abused when mastering engineers attempt to "remix" from a two-track, which is a bad idea. The MLA-4 is not a surgical multiband. This is a mastering engineer's gentle three-band for someone that wants to do large, subtle changes, it's not intended to remix. I can't go in there and fake out the bass drum and make it pump, although

I can do a little of that. It doesn't do "aggressive" very well, and for that I have the API 2500. Or the Pendulum ES-8, both of which do their duty, depending on the needs of the music.

The MLA-4's upward expansion allows me to somewhat "undo" over-compression from a mix. It can enhance the rhythm or the pop of a snare, for example, that might have been pulled down too much by mixing com-pression. But unlike many digital upward expanders, the MLA-4 sounds warm, with its unique analog circuitry.

There's also the AnaMod ATS-1 analog tape simulator, which I think sounds better than any of the plugins—it's got everything you need except the smell of oil and capacitors, and no wow and flutter, which is nice too. The Pendulum Audio ES-8 is a fantastic piece of colour. This can be creamy, or crunchy, or warm. It can't be transparent, though, so if you want transparency, don't go through this, but if you're looking for warm-ing things up, or for doing that silky vari-mu, slow, easy, fattening thing, it's great.

Q: Loudness normalization has taken off recently, to some degree or another. How has this impacted on your practice or what artists/ producers/labels ask you to do?

Many artists are now aware of streaming services being loudness normal-ized. However, until: (1) Apple music and iTunes on the computer and IOS becomes normalized by default; (2) the mysteries of YouTube get answered by Google themselves; (3) Spotify stops peak limiting and starts using such a high target level; (4) Tidal normalizes their MQA service; and (5) Cenelec upgrades their restrictions on mobile devices in Europe, then movement in the loudness area is still stalled, and the situation is still quite frustrating.

In general, my loudness practices have not changed. I always go for the best-sounding master. This means far more than just loudness—it means dynamics, sound quality, stereo separation, warmth, tonality, impact, and many other qualities. If the client complains about the master's loudness, I discuss the normalization situation with them and, after discussion, either raise the master or leave it alone. More often than in previous years, we leave it alone. More often the client takes my word about the loudness. It's a bit less of a struggle than before.

I do pay a bit more attention to the integrated LUFS measurement. That is, I check it. But my monitor gain and the perceived loudness of the album governs my decision whether I should raise the master before sending it to the client, NOT the integrated LUFS measure. I also do not conform to the "lowest common denominator." My nominal loudness target is that of iTunes, −16 LUFS, which is lower than YouTube and lower than Spotify. However, the integrated LUFS is just a guide. In other words, for a popular music recording, I often use a consistent monitor gain of −9 or −10 dB. This is measured relative to 0 dB = 83 dB SPL C-weighted, Slow per channel with −20 dBFS narrow-band RMS pink noise; −9 monitor gain conforms roughly with a "K-14." I sometimes find an integrated album level that reads as low as −17 LUFS but that simply does not worry me if two things are the case:

1. The sound is loud enough to my ears at the −9 monitor gain, especially when mastering acoustic music that has the acoustic advantage. Most of my acoustic clients do not hear an issue in this case—only the clients who actively compare the loudness of their album against other acoustic releases that have been overcompressed.
2. The short-term loudness measures several dB above the integrated programme loudness. This is a very important key to the perceived loudness of the programme. In other words, even if the integrated loudness measures −17 LUFS, if the short-term loudness reads several dB higher than the integrated, this indicates dynamic movement. Short-term loudness is what the ear keys into. This raises our perception of the album loudness, even if there are soft passages as well. We should all credit Ian Shepherd as the first to make this discovery. But I had long ago realized that "something was up" with the integrated measurement, because my own masters, which are often more dynamic than the competition, sound louder than their integrated loudness would seem to indicate. In other words, the short-term to integrated loudness ratio of my material is higher than a lot of the competition, so my material competes quite well, especially after normalization.

The −17 LUFS average figure simply indicates that there are enough soft passages to offset the short-term passages in the integrated measure, not that the album is too low.

Q: What of vinyl's resurgence. Are you cutting more? Are your colleagues cutting more? Are there any startling changes you'd wish to report?

I do my share of vinyl, a very small increase compared with a few years ago, maybe 3% (rough estimate) of all my projects in a year request vinyl, where it used to be less than 1%.

Q: In your earliest days. when did you say, "I'm going to hang a shingle as a mastering engineer"?

First of all, I'm always learning. It would be egotistical and incorrect to say that I've always had the abilities and knowledge that I have now if I look back to 1971, when I started as a recording engineer. The answer is no, I keep learning. I began, and still am, an audiophile recording engineer who records direct to two-track with a minimalist microphone setup. It's that very training with the natural sound of a performance, an orchestra, or a band, or a folk group or a jazz group, that gives me my unique orientation and maybe a heads up on some other engineers. If you look at Doug Sax, who recently, sadly, just passed away, he began when I was still in high school. I listen to the direct-to-disc recordings that Doug made as absolutely the most transparent, beautiful, and impacting records that sounded so much better than other records of the time. This was recording direct to a lathe. So he was right there while that was being done, and I'm sure that he also got an education right then and there, the same way I got an education when I began to work for Chesky Records, recording all those great jazz, and some rock and some pop recordings, direct to two-track with minimalist micing, in great acoustics.

I'm a clarinetist from way back, so I know what acoustic music sounds like. But while recording, I learned what it means to stand in front of a big band, 10 feet behind the conductor, and hear the impact of that band, and I would say, "My God, I could never capture that, but I can certainly try." It's an experience that very few mastering engineers have today, which is, what does real, live, acoustic music sound like? I incorporated that knowledge and experience, just as Doug Sax did from his direct-to-disc experience. I'm sure that it translated to some degree to his mastering of pop music that came to him on analog tape, and eventually on digital sources. I try to apply the live quality of live performances and live music to everything it is that I do, whether it's hip-hop, rock, folk music, or classical. Even with the most artificial of media that are constructed using samples and little tiny pieces edited together in Pro Tools, I find that incorporating the feeling of a live performance helps the master. Then when it comes back and they hear it, they say, "Wow, that's great." I may actually have been thinking about the natural dynamics of a big band while I was working on hip-hop. What I'm trying to say is, it all comes back to the sound of live music.

Kevin Gray

Kevin Gray has mastered music for every major label in every genre, from pop to classical through heavy metal to rap. He has to his credit more than a hundred top 10 and Grammy award-winning records, and dozens of RIAA certified gold and platinum albums and singles. At only 18 years of age, Kevin was already cutting records at Artisan Sound Records in Hollywood, working on releases for the likes of America, Paul Anka, The

Beach Boys, Debby Boone, Donald Byrd, Mac Davis, ELO, The Grateful Dead, Freddie Hubbard, Billy Joel, L.T.D., Manassas, The Osmonds, Kenny Rankin, and Redbone. In 1979, he and his business partner, Doug Sheppard, opened their own mastering facility, The Cutting System, Inc., where he worked until he headed the mastering department at MCA Records, beginning in 1992 (working on records for The Fixx, Musical Youth, Red Rider, Night Ranger, The Who, and many others). From 1984 to 1989, he mastered at LRS in Burbank, working mostly on syndicated radio shows for personalities such as Casey Kasem, Dick Clark, and Rick Dees. He also mastered records for The Beach Boys, Rod Stewart, The Yellow Jackets and Celebration. And in 1989, he helped LRS launch into CD mastering. His old friend from MCA days, reissue guru Steve Hoffman, brought DCC Compact Classics mastering to him, and over the next six years they mastered reissues (for CD and vinyl) for everybody from The Doors, Creedence Clearwater Revival, and Cream, to Elton John, Miles Davis, and Lightnin' Hopkins.

In 1995, Kevin moved to Future Disc Systems in Hollywood (who incidentally purchased his custom mastering system). The next year, Kevin was responsible for the mastering of the MCA Heavy Vinyl Series, which included records by The Who, Buddy Holly, Dave Mason, Buddy Guy, and the "Out of Africa" soundtrack. He and Steve Hoffman continued to work on DCC releases, such as Bonnie Raitt, Jethro Tull, and Jefferson Starship. He also mastered new (digital and analog) major label projects for Wang Chung, Paula Cole, Depeche Mode, Gina G, and Erasure. He cut the Grammy-winning Best Dance Recording in 1999 (Madonna) and in 2000 (Cher).

From 2001 to 2010, Kevin mastered at AcousTech Mastering in Camarillo. There he installed his revamped all-discrete, pure Class-A, transformerless mastering system. Projects ranging from the 30th anniversary reissue of Pink Floyd's "Dark Side of the Moon" to over 130 Blue Note Jazz reissues for Acoustic Sounds and Music Matters were mastered there. In addition, Gold CDs were mastered for Audio Fidelity and Impex and SACDs for Acoustic Sounds. At the end of September 2010, AcousTech closed its doors, and soon after Kevin opened his new facility, Cohearent Audio, LLC. On Nov. 1, 2017, Cohearent Audio completed its seventh full year of mastering.

Q: How did you become a mastering engineer?

Well, I was interested in electronics and in music from a young age, and about the time I hit middle school or junior high school, I figured I'd probably wind up as some kind of radio engineer, because I liked radio, too. I was working to get an amateur radio license at the time. I met a guy who moved into my neighbourhood who had a mastering facility. That was Bob MacLeod, Artisan Sound Recorders in Hollywood, and he asked me if I'd like to go down and see the studio, and I said sure. The very first time I walked in the door I saw the gleaming tape machines and his lathe. He had put up a tape and started cutting, and I said, "Wow, this is what I want to do for a living." It was just like a halcyon moment.

Q: So you were literally thrown into the deep-end, cutting records, and sort of understanding it from the ground up?

Oh, absolutely.

Q: So what kind of things were you set up doing first of all, when you joined up?

Well, before I was even working for Bob full time, I was hanging out there on summer vacations and spring break and that kind of thing, and he had me making the tape copies for International, following all the EQ and level settings notes of everything that he had done for vinyl. So I got fairly familiar with the console and fader moves, and how to make all the level and EQ adjustments. So when he actually hired me, right after graduating from high school, the very first thing he had me doing was cutting 7 inch 45s, because they're a little harder to screw up than an LP. If you blow a side, it's not that big of a deal. So I did that for a couple of weeks after I officially started, and then we got the CBS records recut account, when Hollywood CBS closed their studios. So I was doing replacement lacquers for Chicago, Santana, Simon & Garfunkel, and the like, and I thought I had died and gone to heaven. From there I started working with clients, and within a year I was doing all kinds of stuff with clients.

Q: If you had to try and explain it, how would you answer someone if they asked, "What is mastering?"

Mastering is the sonic version of photo retouching; that's what I tell people. It's the way of making all of the songs on an LP, for example, flow from song to song, making minor adjustments, or maybe not minor, in terms of overall balance, bass, midrange, and treble, maybe some slight compression to bring up some things that need to be louder or to bring down some peak levels, that sort of thing.

Q: What are your views about the talk of equal loudness becoming the norm, and along with the iPod shuffle era, the flow no longer becomes the flow of an album, it becomes the flow of everything?

Right, I hate it. I have been one of the biggest naysayers on this whole industry standard compression thing. I've got a blog on my website about it. Ever since the mid-'90s, CDs became unlistenable for me. I haven't bought a CD that was mastered since the mid-'90s for that reason.

Q: What are the things you first listen to? How do you approach the first stage of mastering?

Well, you may or may not know that I'm kind of in a weird niche. I'm doing mostly reissue work on '50s, '60s, '70s jazz, pop, classical, you name it, but it's mostly reissue and it's for vinyl—that's my niche. I also do new projects on vinyl, including projects for the major labels. Occasionally I do brand new CD projects for independents, but not really in the steady stream commercial venue, because I'm so anti-compression. I know I'm not going to give them what they want.

Q. How is it that you listen? Even on a remaster, for example, what would you listen for and how would you listen?

Let's look at both quickly. On a reissue, I'm trying to stay somewhat true to what the original intent was when it was mastered. So I usually ask for a copy of the vinyl, or if that's unavailable, then an early mastered CD, before all the compression started, so I can kind of get an idea. If it's an album I'm not familiar with, and oftentimes I have the record in my collection, but if not, that's a starting point, because then I can sort of see. Now, I've gone to great lengths to build a very good monitoring system and to have really good acoustics in my mastering room, so that I know what's really happening. If you don't have bass that goes down to 20 Hz, I'd be afraid to be EQing the bottom end—but I do, so I know what I'm doing in that territory, so that would be the first thing if I were doing a reissue.

The other point I was making about the monitoring is I think that monitoring is a little bit different today than it was back in the day that the album was done. Especially if it was sort of the mid-'70s, where everybody was going for Westlake rooms with compression ceilings, and the sound got pretty nasty on some of that stuff, and it's necessary to tame it down a little bit for today's flatter music systems. Particularly since a lot of the stuff I do is for the audiophile market, I try to listen to it in more of an audiophile vein [rather] than playing it back on car stereos or whatever, like we we're doing back in the day.

Now on a new project, my approach would be to try to usually sit down and listen to at least three or four of the songs all the way through, without even touching the equalizers, just to kind of see where it's at. Then I'll go back and start making some minor adjustments, if I think they're necessary, from track to track to balance everything out.

Q: What is it that you've landed with in terms of monitoring?

I built a custom system, it's four-way; I have Dynaudio Midrange and Dome Tweeters, and then I have JBL 12 inch mid bass and JBL 18 inch subwoofers, and the system is three-way amplified. I have a passive crossover between the midrange and tweeter. It's all gentle slopes, it's all very audiophile-type amplification, and the system is flat to well below 20 Hz, actually almost down to 10 Hz.

Q: Is that why you've got the 18 inch as well as the 12 inch?

Yeah, well, the 12s are mid bass, they only go down to 100 cycles. I have actually dual 18 inch subs on each channel, and they go down below 20 Hz, so that's basically my monitoring.

Q: What made you go down the route of actually doing something so unique?

Well I've been building loudspeakers since the '80s. I got into doing this when video first started moving into the home in a serious way, and a lot of people had heard my home stereo, which was custom, and not that different from what I just described to you. So I started building these systems for people, and they were mostly getting used in audio rooms, so at that point I called my company Cohearent Audio and Video, but I've dropped the video since then.

Q: What are you doing about the amplification and the crossovers for that?

It's all custom. I have a disc cutting system that was literally built from scratch, it's not a Neumann system. Granted, it has a Neumann cutting head and lathe, but all of the electronics were built from scratch; it's all discrete, Class A, no ICs, and it's transformerless from input to output. So it's very unique, there's nothing else like it in the world. We basically copied the power amplifier circuit for the monitor amplifiers, with the exception that they're being driven class A/B. The cutter amps only are run class A; it would get a little expensive on the power consumption to run four more class A amplifiers just all the time.

Q: So tell me more about the lathe; what's got you down into the custom area of it?

Well, I only work on my own system; I don't offer services around. There is a guy in New York, Chris Muth, who is also Dangerous Music, he's now using the name Taloowa for his vinyl stuff. He's kind of the main "go-to guy" in this country for parts and services, and he's very good. He started out working for Sterling Sound as their maintenance guy and does all of their maintenance, obviously. My lathe was built back in 1968, and it's a little bit of a chore to keep it running and tuned after all of these years. Now, I have added a Technics SP02 Direct Drive Motor to the lathe, for the Turntable motor, and I have recently incorporated a brand-new disc computer. I was using a Zuma for years, and I worked in conjunction with a couple of other guys, and we developed a new one that we're just getting ready to go to market with. It's being sold under the name DJR Disc Computer, and I'm using one of those. So other than that, the lathe is stock, cutting-wise.

Q: What's your view on the future of vinyl, because there are no more lathes of that calibre being made, and it's a limited market. There's only one company we know of, called GZ, who press a load of things in the Czech Republic, and they're investing in their own presses, but loads of people threw away their machines in the '80s when CDs

started coming around, but now we wish we all had them stockpiled in a hangar somewhere.

Exactly, it's a problem. I wish I had a crystal ball, because I could have never predicted this vinyl resurgence in the first place. I love vinyl, I've always loved vinyl. I have a large vinyl collection, somewhere in the thousands, and I was very happy to see it come back. I was just lucky to have kept all of my equipment for all those years. I started back into vinyl in a big way just shortly after it had pretty much gone out in 1992. DCC Compact Classics, that I was doing a lot of mastering for, and they were doing gold CDs, decided one day that they wanted to start doing vinyl again, too; I thought it was a whacky idea. As far as I was concerned, the door had just been closed on vinyl. So I was really very much at the forefront of this whole resurgence. Then in '95, I did the heavy vinyl series for MCA, which was the first major label to jump back in, and from there it's just taken off exponentially, it's crazy. But you're right, it's a real problem. Right now, we've got a real glut of vinyl, and the turnaround times have gotten crazy, and I'm really afraid that the quality is suffering because of it. I predicted it would happen when the major labels jumped back in, and it has. Leaving it to all of the little labels to do the reissues, everything was great. As soon as they jumped back in, they're trying to get thousands and thousands of records pressed, and there just isn't the capacity.

Q: Material-wise, I hear there's some issues with the materials coming out; even acetates are having a problem at the moment, aren't they?

Well, I've had very good luck with Apollo masters; I've stuck with them, and I really like their product. I've had very few problems, and they've been very good to me, because I buy a lot from them. But yes, I have been hearing that, although I haven't experience that first-hand. I don't know if it's vinyl or other issues in the arena of record pressing, but I've heard a little bit of a downturn in quality in the last year or two, and that's disconcerting. By the way, I ship lacquers all over the world. You mentioned GZ, I have sent stuff to them, Pallas, Optimal Media, Record Industry, MOV; I don't think I've shipped anything to the UK, but all over Europe.

Q: Did you build your room yourself?

I did.

Q: Any particular features worth noting about, anything interesting?

Well, I don't implement this fully in quite the same way, but back in the '80s when they came out with the live end/dead end rooms, I liked the concept of basically having the front of the room fairly dead just to kill first reflections, and then liven up the back of the room so it sounds like a typical listening room, and so that's kind of the approach I take.

One interesting thing that I've discovered is that when people come in here to video, and this has happened quite a bit, they always ask me for the audio that we're playing on a CD or whatever that they can dub in later, and without exception, every videographer has said, "I can't believe how good the sound is in your room picked up from my camera mic, that I've never had to use the audio that you've given me." So anyways, I think that speaks something to the acoustics in the room. I like to keep the back of the room a little bit livelier just so that the overall reverb time isn't too dead.

Q: When you're mastering something, or say the new CD stuff, is there a preferred signal chain that you use, or do you think of it very differently? How do you approach the physical work?

I have my own approach. I basically try to keep analog stuff analog and digital stuff digital. This whole concept of taking digital and going back through analog makes no sense, because I can hear it going through the converters. Even when you're going high-res, I don't like what it does—the fewer conversions the better. So since I'm not into doing all of the multiband compression, I basically use an L2 hardware version and a Weiss EQ, and that's what I use for CD mastering or doing high-res digital. I also do a lot of SACD work, too.

Q: What's your view about high-res? For example, Tim De Paravicini said many years ago, back in the '90s, that until we get to 192 kHz/24 bit we're not going to get close to analog. What's your view on all this?

My opinion of Tim De Paravicini . . . is that he's built a lot of great equipment that I really like. I like his tube EQ, and he does say a lot of things that I do agree with about analog, but every once in a while, I just have to disagree when I read some of the things he's written. We did some tests with a live mic feed when I was working at Future Disc Systems in Hollywood. We had just gotten the Pacific Microsonics Converters, and Sony had just come out with DSD, and we arranged a shootout. We had a live mic feed feeding a jazz quartet from the studio next door, and we had around 10 engineers sitting in a very large mastering room, in the big room at Future Disc, we called it "the cathedral." Anyway, we were listening to the live mic feed that was going through the Pacific Microsonics and the Sony converters for DSD. They were going to an Ampex ATR 100 half-inch IPS, and then we could listen to the outputs of all of these and compare it to the mic feed. Everybody's ears in that room, once we got the Pacific Microsonics, which is a damn good converter, up to 20bit 88.2 kHz, nobody could tell the difference from the mic feed anymore. It was making more of a change going through the Ampex ATR 100, and it was making a hell of a change going through the DSD converter. It was apples and oranges what was going in and what was coming out. With the Pacific Microsonics, 16 bit/44.1 kHz just wasn't cutting it, but as soon as you switched it to 20 bit, that opened it up, then you go up to 82.2 k/20 bit, and that opened it up. We then went up to 24 bit/88.2, 24 bit/96 k, 24 bit/192 k, we weren't hearing any further improvement from a live mic feed. Now you could say maybe we weren't feeding in the right pro-gramme material, but we had splashy little cymbals with brushes, there were lots of nice top-end detail stuff. So I came to the conclusion at that point, and with further listening tests that I've done, that 24 bit/96 k is plenty good enough.

Q: In our experience, it isn't as much about treble but about the depth in the bass—I don't know if you agree with?

Oh absolutely, and that changed more with bits than with frequency; 16 bit at 96 k doesn't sound as good as 24 bit at 44.1.

Q: But again, we've done it, let's say, 48/24 to 96/24, and the differ-ence again in depth and bass, and how that compares to vinyl—it's very interesting the thing you said you did at Future Disc, from the analog Ampex and all of that; I mean, which did you prefer at that stage?

I thought that at that point we were going 24 bit/88.2 k with the Pacific Microsonics, and I thought that, in terms of capturing what was coming in on the mic feed, that that did a perfect job. Anything more is a waste of storage space. That having been said, I love analog, and I love both vinyl and tape, but I think they both do have a sonic signature. I think it is a colouration, but it's one that I like. I mean, I really like the sound of going to half-inch 30 IPS analog; I don't think it sounds exactly like the input, but it's one of my favourite formats, and the same goes for vinyl. But I don't like to get into this argument of which is better, is the vinyl better than a high-res download, and it's like, "Well, what do you like?"

Q: So the loudness wars: what do you hope for the future, given that we've got the European Broadcast Union R128 standard in equal loudness; do you think there's a future that's going to bring us back from the brink?

Well, I don't quite know how to answer that, it's pure guess work, obvi-ously. Here's my take on it: one of the things I love about vinyl, and I think one of the things that the vinyl buyers love about vinyl, is that you're get-ting a better chance of getting an uncompressed version of it. Obviously, there are people who are taking their compressed CDs and putting them onto vinyl, and I hate that, too. I try to ask my clients not to do that. So on most of the stuff that I do from digital, I work with two or three mastering engineers who send me premastered files, and I've gotten them to do a first pass when they master, which has all of their EQ and everything and their level adjustments, but without the overall envelope compression that they're doing for their CD. So that's what they'll send me as a file for cut-ting, and then they go ahead and stomp on it for the CD or mp3, or what-ever digital release other than a high-res. Hopefully, they're using that too for the high-res, but I don't know. Yeah, that's the reason that I think vinyl is still a viable medium, because you can get the uncompressed thing. It seems, in the digital world, that people are not going to change.

Q: They're still going to be expecting the loud stuff coming in, which just seems to be the de rigueur these days?

Yeah, I'm very unhappy about it, but yes. I've been hearing mastering engineers for almost 20 years, when it started, say that they hate this compression, but they keep doing it, and I'm not one of them and I'm still in business. I've got this vinyl niche, and it's been very good to me, and that's probably why I'm still in business—otherwise I'd be out selling real estate or something else, I don't know, because I hate compression so much. One of the reasons I left Future Disc Systems is because I was mostly working on rap and dance music, it was all very compressed, and I was asked by Record Technology and AcousTech Mastering to go up there and do vinyl reissues of stuff that I loved, so I thought, "Well, okay, I've got a choice of this and that. Okay, I'll take that one."

Q: How has the mastering industry changed since you've started?

Well, I was kind of on the forefront of that, too, because vinyl reigned supreme when I got involved, and by 1986 everybody wanted CDs. So from '86 to '92, all I did was CD mastering, and of course that was back with the Sony 1630, and we can talk about the dinosaur realm of that, too. However, the heavy compression thing to me didn't start until about '95, that's when guys like Steve Marcussen and Eddy Schreyer on the West Coast, and Ted Jenson and all the boys at Sterling, I don't know where in the UK it all started, but those were the guys, and Vlado Meller, they just started squashing everything to death, and I was like, "What the hell is going on? Even with 16 bits you can get really nice dynamic range, and we're squashing it into 8 bits!" Then 24 bit came along, and I said, "Well, what do we need 24 bit for when everyone is squashing it into 8 bits?" It's ridiculous.

Q: If you were starting out on your mastering journey now as a young-ster, what would be the one thing you would say to yourself?

Where can I find a gig where I don't have to squash as much? Seriously, I think that would lead me into trying to do jazz or classical mastering as opposed to pop; I just find it offensive. I put on a CD mastered after 1995 and I just can't get past cut 3 at any volume level. It's not just listening to it loud, I just find it really irritating. I'll get off my soapbox now.

Q: Could you talk us through the process of how you work with your lathe for a reissue and how you would start your project?

As I mentioned, generally I ask to get a vinyl pressing or an uncompressed CD, meaning one that was mastered before '95, and I start out listening to that. If they send it to me soon enough, I'll take it home and listen to it, just so I can get myself familiar with what the sound was that they were going for. In the old days, a lot of the mastering notes were in the tape, but over the years they've disappeared on a lot of projects. So that's my starting point, and then I can bring the pressing in here and A/B it against the tape or the high-res digital file, hopefully, or whatever they send me, and try to get it in that ballpark. That is the primary goal. The secondary goal is, "Okay, can I make some minor improvements?" Say, for instance, they filtered all the bass off below 40 cycles. I know that in the day that was usually to keep it from skipping on cheap record players, but we don't have that issue quite so much anymore, and sometimes it was compressed for the same reason. If you go back to the mid- to late '60s, a lot of this stuff had a big midrange boost just to make it cut through on radio, even on the album, not just on the single. So I might tone it down a little bit from where they put it there, but the thing is, so many people buying these records want that vintage sound, they're used to that sound, and they don't want it to be an apples and oranges difference, they just don't. In a few mastering situations I've tried to take a record in a different direction, and I've gotten nailed for it every single time, so I've learned from that. So I might tone things down in the midrange if I think it's really bright, and I might let the bass go a little bit deeper, and I might not compress it quite as much, or at all, if I can tell they've compressed it. That would be a given, that would be the main thing I wouldn't feel too bad about changing, but that's my basic approach.

Q: So what's the analog chain you use to get the audio from, say, the tape through to your lathe?

I have two Studer transports, and they're transports only, they have custom electronics and heads. I'm using Flux Magnetic heads, and the custom electronics are basically pretty similar to what's there, but in a discrete form rather than with ICs, and maybe a fewer stages here and there, and a few other details other than that. I have a half-inch that's configurable either as a half-inch two-track, I also have three-track heads for it with an extra set of electronics, and so I've actually done some records from like '59 through '62 that were done three-track, direct to the cutting lathe through a mixer. Then on my quarter-inch machine, in addition to Class A solid state, I also have Studer C37 playback electronics. So if people are really going for that tube sound, I can give that to them, too. I love those electronics; I have a full C37 recorder also and plan very shortly to start doing some direct to two-track jazz with that.

Q: Do you wonder if there's going to be an open reel revival shortly?

Well we've seen it in this country already, maybe not in the UK yet, but there's a label called "The Tape Project" that's releasing 15 IPS CCIR tape copies of major albums, just like the vinyl guys are doing the reissues of. They're charging $450 for two reels, they have a line of machines, but they're duped in real time, they aren't done high speed; they make a 1-inch two-track master and then they dub them from that. I think the whole thing is a little ridiculous, but for those guys that really want tape and have the money, go for it.

Q: So you've got the tape machines with the custom electronics; how would you do the processing from there on in, as you transfer to the lathe?

My chain is also completely custom in the console, my own line amplifiers Class A discrete, and no transformers as I mentioned. We have custom equalizers that are based on Massenburg's early Sontec circuit but using our own line amplifiers for the output stage on that, and I have some custom limiters which are basically UREI 1176s, but the only part of it that's actually 1176 is the actual gain reduction FET circuit and the side-chain that drives it. The output amplifier, the gain makeup amplifier is one of our Class A line amps, too, and no transformers. So I have two sets of EQs and level controls and limiters, and that's my basic back-and-forth chain from track to track on an LP, for instance. I can crossfade between two settings and that sort of thing.

Q: Is there any way we could get a picture of your console?

There are pictures on my website: www.cohearent.com in high-res. Me, Bernie Grundman, The Mastering Lab, we all built custom consoles. Most other people have got modified Neumanns that I've talked to; there really aren't many custom consoles out there.

Jonathan Wyner

Jonathan Wyner is a chief engineer at M Works Mastering, technologist and education director for iZotope in Cambridge, MA, and professor at Berklee College of Music in Boston. A musician and performer, he's mastered and produced more than 5,000 recordings during the last 30+ years. Credits include James Taylor, David Bowie, Aerosmith, Kiri Te Kanawa, Aimee Mann, London Symphony, Miles Davis, Semisonic, Thelonius Monk, Pink Floyd, Cream, Bruce Springsteen, and Nirvana.

He has several accolades, including production of the Grammy-nominated soundtrack for the PBS special *Invention and Alchemy* (Deborah Henson-Conant, 2005), the mastering of the first recording of a full-length opera (*Madame Butterfly* 1912, BBC), and the first interactive CD game (Play it By Ear, Rykodisc). In 2012, he authored *Audio Mastering: Essential Practices*, published by Hal Leonard/Berklee Press.

Jonathan lectures on music and audio topics around the world, having presented with George Massenburg, Doug Sax, and Bob Ludwig among others. With the iZotope team, he authored and developed Pro Audio Essentials, an online ear training site that garnered a TEC award.

Jonathan hosts *Headroom*, an online webcast series archived on iZotope's website.

Q: How did you get involved in mastering?

Well the shortest version possible is that I started out as a musician and was completely enamoured and taken with the idea of marrying music and technology, which I did in some unconventional ways using horns and synths in the 1970s and early '80s. I then got involved in recording and producing and was just starting to develop a career when I got a gig in a mastering studio. . . . I was really taken with the idea that I could actually work decent hours, and the quality of what I was working on suddenly skyrocketed, as I went from working on demos and things that *might* become to working in mastering where, at least in 1985, CD releases meant something in terms of quality, so that was kind of the genesis of it for me.

Part of what was interesting to me about getting involved in mastering, especially at that time, was that digital audio was just becoming a "thing"

70

in the wider marketplace. It seemed like there was a lot of magic to it and a lot to be learned, and there was a great opportunity to dive head first into the nerdy side of my interest in music technology, and it was really just an accident that I was offered the gig, sort of right place at the right time. That turned into an opportunity to work on lots of fascinating records with some very well-known artists that I may never have otherwise if it had been a different time or different place.

I think I was very well suited to being a mastering engineer in a couple of respects. One is that I grew up in a rich musical environment—there was classical music at home, composers in my family generations before me, and I was surrounded by pop culture and radio playing the beginnings of the maturation of rock & roll—I was listening to everything and anything, and I had a very wide musical palate, and I think as a mastering engineer that's very helpful, to be able to be informed about various styles. One of the things I'm always fascinated by is how there are certain audio and musical issues that cut across genres, and while we might talk about level, or persistence of level, in pop differently from classical music, we're still talking about the same thing, we're talking about physics of sound, so you start to notice similarities across genres. So being versatile musically I think was helpful, and being interested in the technology and the marriage of the two . . . and also being patient.

Q: I suppose with those early 1610 machines you would have to be very patient?

Well, yeah. It was both a blessing and a curse. Everything was slow from back then. The 1610, it was a tape based format, you had to record all the way through, and then you had to rewind the thing to listen back, so with the early tape based systems you had to listen to every record at a minimum of three times without even thinking about the actual mastering work.

Q: Quality control was happening for sure in those days as a result?

Yes, it was. Unfortunately, QC is a dying "art." I teach at Berklee College of Music as well, and one of the things I talk to my students about is the art of slowing down time and listening and observing as part of the whole QC process.

Q: If someone asks you what you do, how do you describe mastering to them?

Well, it's sort of a weird metaphor, but in some ways I feel like it's being a midwife. Someone's about to bring a creation of their imagination into the world, and I do my best to shepherd it. It's maybe an imperfect metaphor because midwives are dealing with the beginning of life, and then mastering is a little bit more about something becoming mature into the world. I think mastering engineers are assistants; we help curate and hopefully *help* to realize the creative vision, and then to make sure that something goes out into the world and can be consumed by people well.

So the classic definition is the final step in production and the first step in distribution, and we're crossing that bridge, and everything that comes along with it. I think you have to inhabit both the aesthetic and the technical sides equally. I think you have to be sensitive and able to execute on the aesthetic side, whether it means strong intervention or light or no intervention. If something comes in sounding perfect, why in the world would we mess it up? But on the back end, data has to be formatted, bits have to be in their places, sample rates have to be set properly, any known distortions have to be logged and accounted for, and you have to make sure you're not introducing anything new, and then it goes out into the world and you keep your fingers crossed.

Q: How did you learn to master, and who were your teachers?

Good question. I started working at a facility called Northeastern Digital Recording, and the man who started that business was Dr. Toby Mountain; he was a doctorate. I can't remember exactly what his degree was at UC Berklee in California, but he wasn't an engineer first and foremost, he was a musicologist. He had completely embraced digital audio technology and started a business to address this new platform, which was the CD.

He was a very bright man and figured out a lot of things about how to work the technology, so I spent my first maybe 12–18 months working for him just as the quality control engineer. I learned a lot about hooking the gear up, I learned a lot about what could go wrong, and I spent a lot of time listening to records. So my teachers in a way were all of the mixing engineers, and even some of the mastering engineers, who would be sending in material that I would be checking or compiling.

So, I got to listen. I did a lot of listening, and that's how I learned. It's a very different thing to go from that passive mode to the doing stage, and I remember my very first session, in which I was suddenly terrified because I realized that now the time had come to move from map point A to the result of point B, and that was a very different experience, but I could lean into all of that listening that I had done. There's something about what a record sounds like, in terms of balance, frequency distribution, and level, and I had educated my ears an extent by that point. I had all that listening experience to lean into, and I think that's something you only get by repetition.

Q: What are the things that you typically listen for when you receive the music to be mastered, irrespective of genre? What's the thing you focus on?

It is difficult into to break it down and become overly specific. You want to lead off with statements like: you listen for balance; you listen for overall tone; you listen for a sense of dynamics in the music; you listen for transitions from quiet to loud passages; you listen for "intelligibility." A well-recorded, well-mixed record is one where the listener always knows where to focus. The listener always knows where to focus their attention, and so when you're listening to a record. the question is, "How much work do I have to do as the listener to have access to what it is that I feel like I'm supposed to have access to in the moment?" In order to understand that, you have to understand something about style. There are all of those issues that are musical-aesthetic issues, and musical arrangement issues, and then there are the obvious things like, "Oh my gosh, the kick drum transient is 10 dB too loud, and this is a record that needs to be loud in the marketplace. What to do!?" So there's a technical issue I'm going to need to deal with right there. Or there's something that's hurting me in the upper midrange, or in the presence region. You're logging notes about musical items, musical issues, and also technical issues all at once. As a mastering engineer you inevitably have to notice the level coming in and think about the level going out, because that's part of the journey, and you incorporate that understanding to develop a sense of what your strategy might need to be in order to build the bridge.

Q: So how do you approach it in the sense of a preferred signal chain? What do you generally reach for?

Firstly. I'm not a product guy [despite involvement with iZotope]. There's no requirement for me to be a product guy here, which is one of the reasons that I'm well suited for this job [at iZotope], so I don't mind talking about any tools.

I know it sounds a little bit like a cop-out, but it's not meant to be when I say it really depends on the programme material. If I have something where I think a very, very light hand is what's required, then it's one set of tools, and I might be looking at working entirely in the box, or at least staying all in DSP. If I need a little bit more intervention of a certain sort, I might need to come out into my analog desk. I will say I have some favourite outboard DSP boxes like the Weiss boxes—I still reach for the EQ1, occasionally the DS-1 and occasionally the TC 6000. They're old friends, and they have their uses, although I've been leaning more into software based limiters when I need them. Then in terms of "in the box," I have some favourite software tools from iZotope, and DMG; those are probably my two go-to toolsets. But if I need analog, right now I'm using the Merging HAPI convertors to wrap around my analog gear, and I'm extremely happy with it to the point where I sold my Pacific Microsonics because I could. Finally, I'd gotten to the point where the conversion was at least as good if not better than and more reliable.

Then I've got a combination of new analog and old analog in my desk. I've got an old Sontec mastering EQ, I've got some of the newer SPL pieces which have all of the headroom in the universe and are very low noise; I'm very fond of those. So the two chains are either all DSP, maybe all in the box, or some DSP in front of a DAC for some corrective measures, analog gear, recapture, and then maybe a little bit more DSP on the back end, whether it's just limiting or a touch more EQ and limiting, and I always keep an awesome reverb around just in case, because every so often reverb does come into play in mastering, and people are often surprised to hear that.

Q: Now an impossible question, our apologies in advance—are there specific EQ moves that you tend to find yourself always reacting to today rather than in times gone by?

The problem that I have with the premise of the question is that there is an assumption that there is a single version of a particular type of problem. So, to sort of pull it apart a little bit—yes, mixes on average are not as well balanced as they used to be. Probably the biggest problem is in the low end, but the problem isn't always that the kick drum is too loud and the bass is too quiet, or that the bass is too loud and the kick is too quiet. It's just that there is a problem. So getting the low end, especially in the bottom two octaves of the spectrum, to sit properly in support of the rest of the spectrum is a huge issue in order to make a record that sounds well balanced and wide and full and so on. So I do have strategies that I use for dealing with one or the other version of that problem, which usually involves some combination of either some low band compression—if you think about a compressor as a tool that you can use as a transient designer, then it's a way to either encourage the tonal base or discourage the transient in the low end—and then sometimes that will fix the problem; sometimes I'll have to use instead a combination of EQ moves in the bass where I might be boosting a little mid bass and shelving off the very deep bass below 40–30 Hz, getting rid of some of the noisy, mechanical sound, or some combination of both strategies. So that's kind of one typical kind of problem that shows up in mixes, but the problem isn't the same every time.

Q: What we've been finding is that lots of engineers are saying they're finding these issues, and bass is one of them, but others have said the high frequencies are very cluttered and respond differently.

That's the other thing I wanted to say. Especially with the advent of the look-ahead limiter, and now easy access to multiple instances of limiting and the non-harmonic distortion that limiters generate, the accumulation of energy in the top end is amazing. This is one of the things that's different now from when I started—once I'm done with a record, it's often darker than when it started, and back in the day I think it used to be brighter than when we started. It's a funny thing; I hadn't really thought about it before in that way, but I think it's true—it's so easy to over-process, and the build-up of the distortions in people's sessions, they accumulate in small increments, but it's really noticeable at the end of the mix.

Q: Do you do any de-noising or editing in your work?

Oh, yeah. So, I use de-noising tools in every session, there's always something to do. But I've done a bunch of remastering; in fact I worked on the entire Bowie back catalogue for Rykodisc, from "The Man Who Sold The World" through to "Scary Monsters." That was back in the day when

"NoNoise" was the only thing around, and so 5 minutes of stereo audio would cost $10,000 to process, and the systems were $25,000 USD each, etc., so it wasn't a trivial thing to undertake.

I've worked on things for the US defence department, some forensic work; I've worked on and remastered the very first recorded full length opera, which was in the national archive of New Zealand; it was *Madame Butterfly* recorded in 1912 by the BBC, so I've done a lot of that remastering work as well, for classic blues albums, Sun Records, and so on.

In short, I've done a lot of it.

Q: So you use it all the time, then?

I do, every day, in every record. I think that one description of the job of a mastering engineer is to make sure that whatever is in the music that should be available to listeners is available, and that anything that gets in the way is not. So whether it be a loud click, or a mouth smack, I mean some mouth smacks are good, but some we want to get rid of, or guitar squeaks, or even occasionally compression holes, or bad edits, and we can use de-noising tools to remove the artefacts and therefore make the whole listening experience better for the user.

Q: So lots of use of iZotope RX then?

Lots.

Q: So what's your feeling on mid/side and how often do you use it?

Well let's see, I use mid/side every time I play an LP, I use M/S every time I listen to an FM broadcast, I use M/S every time I make an mp3. I'm not trying to be coy, but I'm pointing out that mid/side is one of the two ways in which we can describe stereo, and so we're always thinking about it and thinking in terms of M/S; it's one of my favourite making techniques.

In the context of mastering, it's an invaluable toolset that gets used very occasionally when needed. There are some obvious problems that come into play when you use mid/side having to do with phase relationships and the impact on stereo image and the soundstage. If you're going to go to mid/side, you have to want to go there because you've got an issue that you need to deal with, because there's going to be a little penalty. Having said that, in mid/side mode, if I'm just de-essing the mid and I don't have to mess the sides up in certain instances, it's genius, it's brilliant. Every so often to create a slightly greater sense of space and to declutter a mix a tiny bit, fantastic. Mid/side compression is a nightmare; I see people using it and I'm like, "How is that going to work?"

Q: Especially multiband compression in mid/side?

Well yes, but it sort of doesn't matter, because if the centre of a mix is moving differently from the sides, and in one moment you've got some present dry signal, and then suddenly it becomes soupy and you lose clarity in the middle, and in the correlated signal.

You know, with great power comes great responsibility, right? You sort of have to understand what the implications are of wandering into these waters.

Q: Are there any particular tracks that you've done or worked on where you've really found that M/S is the only way to have fixed that record, without having to go back to the mix engineer?

Yes, of course. I think probably the kind of problem that I'll encounter that causes me to go into M/S mode, and apart from the specific issues where I need a little more kick in the centre or a little less "s" in the centre, is when someone's got a monitoring environment that's too narrow, and they'll increase the energy, especially in the low-midrange.

I was working on a record where there was an acoustic guitar panned wide, lead vocal in the centre, and there was way too much "bloom" in the low mid from the guitar, and it was covering the vocal. The only way to address it was to pull a little bit of low midrange energy out of the sides. I could have done it with a stereo tool, but then the vocal would have ended up being too thin. So that's the kind of issue, and it usually results from poor monitoring, or slightly miscalibrated monitoring. In those instances, it's exactly what you need to do.

Q: Loudness is obviously going to come up a bit. Besides compression, what are the other ways you might achieve loudness?

If I'm a mix engineer, turning stuff off!

My relationship with the whole topic of loudness—on one hand I'm so completely put off by the whole "loudness wars" discussion, and I'm fascinated by the idea that a well-proportioned, well-balanced mix will sound louder, and a good arrangement will just naturally sound louder than a bad arrangement, or a cluttered arrangement, or a mix that doesn't distribute the energy well across the spectrum. If you think about ISO226, the new version of Fletcher-Munson, if you incorporate that understanding into your work, you can certainly achieve a little bit more level by cheating out a little bit below 40 Hz and add a tiny bit around 2 kHz. A half a dB in either place, and you've got more than a dB of difference in terms of perceived level.

Not pushing too much deep bottom, or too much ultra-top into anything. If you want an impactful sound through the midrange, you need to focus on the midrange. So the boundaries in terms of maximizing loudness, and I'm just pulling some numbers out of the air, but maybe between 100 Hz and 5 kHz, that's sort of the area where we're wired to be more sensitive. There's an associated issue to that which is the consumers' playback systems are not always going to represent spectrum that lives outside of those bounds. I mean this phone [pointing towards his cell phone] isn't going to give you lower than 250 Hz, so if you want something where there's a persistence of sound coming from the speaker; you don't want to have a lot of the energy at any given moment coming from below 100 Hz at the expense of the midrange. That's not to say that treating 30 Hz or 40 Hz signals properly isn't important; I believe it is from the stand point of . . . if you do, it makes great music and great-sounding records; it's just about keeping things in the right proportion and being able to hear them so that you can do that.

Q: To talk about the loudness wars a bit more, Bob Katz's K-System could be said to have been more or less been adopted in one form or another. Do you think the "loudness wars" are their way to being over?

Oh, absolutely. As much as I love Bob, and with all due respect, I don't think we mastering engineers have had anything to do with it. The loudness wars, as they have been construed, have been a response to technological developments and music distribution. The reason that they're changing now is for the very same reason—that music is now being distributed in a way that is similar to broadcast as it used to be. So loudness normalization is now the word of the day. We've got other qualitative issues that we have to deal with, around the way the compression is dealt with, and how the loudness normalization is dealt with, and so on. But I think that things have changed because of changes in technology. I think the look-ahead limiter allowed us to serve up more level on a meter with less distortion than before. If you look at records mastered before 1992 and look at what happened starting in '92 until about 2002, you'll see this enormous change in average level—what we're now calling the LUFS, and also changes in the low end. I mean the aesthetics of records have changed as a result. If you listen to hip-hop records in the '80s, you don't have as much bass in them because you couldn't, because it produced way too much distortion and the production tools wouldn't support the low frequency sustain. It's fascinating to think about it.

Anyway, yes, I do think things are changing, and I'm happy about it. Right now it's still a little bit messy, because you still have to make a master that will satisfy the fragile ego of the artist when they're comparing their record to some other record and also make something that will sound half-way decent when it gets played on a loudness normalized platform, but I'm happy to have the problem and be thinking about it.

Q: Therefore, do you have to produce many versions of the same master, and is it frustrating?

Is it frustrating? Well, it's challenging. It's not always easy. It's hard to serve multiple masters.

Here's my approach, because it is a thorny problem having to generate more than one version of a record. I mean we used to have to make one for cassette distribution, and one for LP, and so on. This is more different in many respects—making something for peak normalized consumption versus loudness normalized.

So my answer to that right now has been in my first pass, I aim for a level that will survive the transition when it needs to get pushed up higher or it needs to get turned down a little bit. Then from that point I will make two versions of a record: I will make one that is more dynamic and one that is less. So I'm generally generating two versions, one where the average level might live somewhere between −14, −12 something like that, and one for the inevitable request where the artist says, "That sounds great, now can you make it 6 dB louder." You have to prepare for that. It forces you to maybe squeeze the average level a little bit more than what's in the mix, pretty much in every instance, but you don't have to go nuts anymore. It used to be that you knew your target was close to the roof, and so that was the battle we were fighting, but at least now you can give an artist two versions and say, "So this is the one that you can use to send to A&R, or you can use to satisfy yourself to compare against the 'whatever.' You also have this other version of the record that you can send to streaming platforms, and it's going to sound better."

There is education involved, and mastering engineers I think historically have been resources and have had to be educators back to the rest of the production chain about what's going to happen to your record when it's listened to. If you think about LP cutting, which is not something I did personally, but we all understand the issues around highly out-of-phase, very low frequency signals or too much sibilance, and your mix is going to dry up a little bit some of the time anyway when you go to vinyl, so there's some education there, too.

Q: How has mastering changed since you got started, especially in the last couple of years?

Well, the economy around has changed for sure, and it's much faster. Even in the last two years, with 64 bit operating systems, and incredibly large storage devices, and sending things via DropBox or what have you, everything's faster. Some of that is wonderful; the ability to get feedback quickly and to send files back and forth is just great. The problem is, both because of the erosion of the economy around mastering and also because it's so easy to move quickly through things, there's a seduction to sort of do things fast. As I said earlier about slowing down time—and QA and QC—requires a lot of discipline, to sort of make you say, "No, I'm going to do that extra pass. I'm going to take five minutes and then come back to it and spend the time that I've always been accustomed to spending on a project, and not cheat the process." So I think that just in terms of workflow and the ecosystem around mastering, I think those are some comments, leaving the whole level issue aside.

I do think that DSP based tools are getting better and better, and that's something that is really fantastic. I think that we've reached this wonderful place where convertors are excellent, and we have the choice of incredible control and every filter type that we might ever need or want, and some interesting plugins that are creating non-linear processing in a way that's compelling and new and interesting.

It's fun; I mean, I love technology, I love the new possibility that it offers, so I'm excited about the creative possibilities.

Q: You mentioned Merging Technologies HAPI, as that's the interface you're on. Are you using Pyramix or another DAW? And what monitoring do you use?

My choice of DAW is Sequoia

Regarding monitoring, it depends on where I am. At home I have some Eggleston's, but I don't do much work at home. I still use Dunlavy SC-Vs in my mastering room; we've also got some PMC's in the iZotope critical listening room, and then at Berklee we've got the ATC 150s.

Q: How did you get involved with iZotope, then?

Well, it's a funny story. I got wind of the first version of Ozone, and my mastering facility was seven blocks or something from the iZotope office, and I looked at what I was reading, and my first response was this sort of knee-jerk, "Well, you can't do that, you've got to do it this way, it's got to be out of the box, it's got to be whatever." This was back in the day when native real-time processing was still a challenging thing. Of course, I was working with Sonic Solutions and there was onboard EQ from more or less the get-go, but that required a whole separate sort of DSP inside a computer to run a card.

So I looked at the address, and I said, "These guys are close by, I'm going to call them up." So I just picked up the phone and I called them and I introduced myself, and must have said something like, "I have some questions about what you're doing," trying to be political about it. The response was completely disarming, which was, "Oh great! Can we take you to lunch and we want you to tell us everything we're doing wrong." I thought, "This is great," and plus they like sushi and I like sushi, so we had a great lunch and that was the beginning of the relationship.

I spent some time working on issues around the Boston music community, sort of separate from iZotope itself, so I developed a professional relationship with them, and over time. I became more and more involved with the company. Initially, I did some writing, authoring the mastering guide for them. I then did some presentations at various trade shows with them, and then did some more consulting around UI and product development, and then helped select a new space and helped build out the new facility. In addition, I support a bridge to education. The iZotope software is incredible as a teaching resource. Anyway, it sort of all turned into a title and a role, but it's been a relationship that's extended over more than a decade.

Q: Could you detail the development of Ozone?

The impetus for developing Ozone, I think at least to hear Mark Ethier talk about it, is that he was an engineering major, but also a music minor from MIT. He tried to record a senior recital, then listened to it and thought it sounded absolutely awful. Someone said to him, "well it just needs to be mastered." It turns out that the recording was simply terrible and it needed more than mastering, but at the time it piqued his interest, and he thought, "Well, all right, so let me try to understand what that means and what that is." The more and more he read about mastering, the more he became interested, and he and his partner, Jeremy, who were two founders of the company, set to seeing if they couldn't figure out what mastering was and figure out a way, leveraging what they knew about programming and DSP, to build a tool to do it. I think they were met at the time with a lot of push-back like, "Well, you can't do this natively in a computer." Computers aren't powerful enough, and all of that resistance that I mentioned earlier, but to their credit, they didn't listen to that, they relied on their innovation and enthusiasm for it and persisted.

I think that was really the original impetus for why it was developed. I think that there was also the sense that it wasn't going to be long before computers would be up to the task, so let's try to understand the discipline, build some tools, and as computers evolve, the tools will evolve, so I think they were right.

Q: From iZotope's perspective, what was their impetus to continue and change that philosophy or that interpretation of that product [getting accepted by the professional marketplace]?

I think over time the company has become smarter and better informed. One of the reasons why I became involved in an advisory role was to say, "Look, you really need to improve the filters if you want to be taken seriously, and to create tools that are truly useful. You need to pull the delays out of the imager, because you're encouraging people to completely destroy their audio"—to suggest improvements that would make sense not only for professionals but helping people make better music, better-sounding records. I think it has really been driven by the desire to innovate, not just to do new things, but to get better at doing what you do, so I think that infuses the work of the company.

Q: I perceive Ozone Advanced really signalling a change in perception to the product from the professional market. Is that the case, and was it the rationale?

I guess that that's true. I think that in response to that, and I share your commentary, if you have a window that's not resizable with all this stuff in your way, wouldn't it be better to just be looking at one module at a time? Whether or not that was about being more professional or less, no. Of being responsive to what people want, absolutely.

I'll tell you one of my favourite things about Ozone is that it's a stand-alone app, and you can pull other plug-ins into it. So if you've got one track without a lot of editing required, and it's top and tailed, you can actually work entirely in that environment without pulling it into a DAW. There's this variety of different experiences available to people, and I think that's very smart.

It's a complicated subject, because impressions about how good something is, especially in the world of software, depend on things more than simply the performance of the dsp code itself, but also the UI, the way people interface with it. So when iZotope went from Ozone 5 to 6, there were a number of people who commented on the difference in the sound of the EQ. The fact of the matter is that the DSP didn't change, but the UI did, and as a result I think people were getting different results. It's like if you change something about the strings on your guitar—it's the same thing, it's the same guitar—maybe again it's not a perfect analogy. If you think about what we do as a performance, and then you change one facet of what the feedback is, you might perform slightly differently.

Q: That's very true; the new GUI was a huge shift forward in terms of usability.

Yes, and I think that means that you hopefully get better results. There's so much that goes into usability and performance and perceived quality, and that to me is a fascinating topic.

The ultimate goal would be to build products that allow people who perform at the highest level to have the control and achieve whatever it is that they want and **ALSO** make it accessible to a wider audience, an ever increasing audience . . . but there's an existential issue for anyone that makes things, especially software. I'm just going to pull numbers out of the air just to make a point, but let's say there were 1,000 working professionals in professional audio in the world, and there are 1,000,000 people who aspire to make music at home for themselves, and you start to do the math, there's a huge appetite among a huge swath of the population of the planet to do these things. Who are you making things for? It's not possible to be all things to all people, but you have to think about how you can increase accessibility and availability. I think about courses and online courses and Coursera, it's sort of the same dynamic—there's a huge appetite for knowledge and information in the world. Ozone is accessible to a huge audience, where analog equipment isn't, and Ozone helps people make better music.

Q: What are the key aspects of Ozone that you use or you would use more than others, such as the limiter?

Yeah, the limiter and the EQ from the beginning have been the two go-to modules. I might say that those are the two most commonly used tools for mastering engineers, anyway. If we're doing any kind of RMS compression, it's usually light handed, but EQ is something like 90% of the work, and getting a level right is the other 8%, and the other 2% is something else.

So I think those are really the two main tools, and of course having dithering options and a bitscope. Some of the visualizations, I would include those as go-to tools. The spectrum behind the EQ controls, and my understanding is that iZotope was the first to overlay one on top of the other, is another great feature. Having that relationship available just makes a ton of sense, as opposed to having to look over at an RTA (real-time analyzer) and your EQ is over here. So I think those are my main two, and everything else is available when and if needed.

There's a new module coming out called Tonal Balance Control which I think is very exciting and relies on interplugin communication to let people navigate complex problems more simply, and the dynamic EQ, I might mention, too. There's a new tool that's going to help people with spectral shaping in a way that I think might be really useful and interesting.

Q: Let's speak about the modules that were removed. You mentioned earlier about having an attractive reverb in your studio. Could you discuss why the reverb module was removed from iZotope's perspective? The benefit of it was the ability to feed a small bandwidth to the algorithm for adding reverb.

I think that the fact of the matter is, and it feels a little bit risky to say this, but I'm going to say it anyway, the reverb algorithm wasn't very good. The point was, if we were committed to build great tools and again, give people something of the highest quality, this wasn't it. So wouldn't it be better to have something that was better?

The reason that it feels risky to say that is that some people don't like to have their toys taken away. I will point out that you can still pull in the Ozone 5 module if you have the component version of Ozone, but there is a dear friend of mine down in Uruguay, a fantastic Cesar Lamschtein who records with orchestras, and he loves the reverb and he says it just matches his room perfectly, so far be it from me to say it's not a good-sounding reverb. I have used it very seldom, but do occasionally to add just a tiny bit of something when it's been the right colour, but it really just wasn't ultimately all of what you need as a mastering reverb, and so that was really the impetus for removing it from the suite. We didn't have the resources at the time to push the development forward and make it what we wanted it to be, so it was a qualitative issue.

Q: At our Innovation In Music conference recently, iZotope's CTO, Jonathan Bailey, was the technical keynote speaker, and he outlined innovative new ways the company is heading. Where do you see the current machine learning and artificial intelligence ideas going into Ozone at the moment, where do you see that heading in terms of future revisions and how we interact with the software?

Well I think ultimately machine learning has the potential to help us make interesting and better-informed choices about what we do in our work, if a machine is smart, in the same way that a spectrograph is smarter than an RTA or a real-time spectral display. So we can get a more nuanced bit of feedback about what we're hearing by looking at tools informed by AI. If machines can help us understand things about spectrum and about dynamics, and about how we will be able to map audio from one system or playback system to another, whatever it is, I think it has the potential to allow us to make better choices by giving us more information, and amalgamated in a way that's more meaningful, if that make sense. It's one thing to understand spectrum and dynamics individually, but to understand crest across different parts of the spectrum is a much more musical construct.

So in that same way, I think machine learning has the possibility to put multiple parameters together and help us understand what we're working with better and help make better choices.

I think that the thing that's most interesting to me to think about is whether machine learning . . . because it can help us understand models that exist and help us work towards a standard distribution of energy across the spectrum, for instance, —it also has the possibility of causing all of our work to converge and become more similar.

So the question that's out there for me is, in what way can we employ machine learning tools that allow us to be creative, or do things that are different or outside the norm? I'm encouraged as I look back through history; if you look at MIDI, for instance, where everybody was upset about MIDI because string players were going to be out of work. Well, the fact of the matter is that MIDI has allowed us to do a lot of things creatively that we couldn't do before, and that in part is because people would dream up their own ways of using it. So, my hope is that any machine learning-informed tool would allow people to sort of react against or interact with it in ways that we can't even imagine and do creative and interesting things.

The danger for me is that, sort of that tendency towards sameness. If the machine says put the thing here because that's where it's supposed to be, and it's smart and well informed, and if we're talking to users who are not as experienced, are they not simply going to do that thing? We have a lot of conversations about that, and we work very hard to think about developing tools that help people be creative, as opposed to automating the process. Ultimately these are creative endeavours, so why would they need to be automated?

Q: Neutron must be selling like hot cakes, I imagine, at the moment, and there's a level of automation there that is very well received.

I think the EQ learn function is amazing, and actually it doesn't even do anything for you other than set EQ nodes. I think the technology is fantastic. The track assistant is an interesting tool to use as a guide. We give people tools and watch what happens.

Q: There's a new piece of kit that iZotope is bringing out called Spire?

I love it, it's great. It's a very high quality either mono mic pick-up or two-line input device. The UX is fantastic. You can quickly sketch up to eight tracks of ideas, and you can mix them with this little interface on the phone, then you can push it out as a multi-track or a stereo bounce out to the cloud, and people can pull it into your DAW. We've already had people say, "We're so excited that the vocalist will stop recording their vocal tracks into their laptop mic, and they can use this instead." Mike Grace consulted on the design of the mic-pre; we've done everything we can to create an excellent quality device. Due to the small form factor, there's been some design challenges, and the team worked really hard to surmount the challenges that having signals in such close proximity brought.

Q: If you were going back to the beginning of your time as a mastering engineer, what would you tell yourself now?

I would tell myself to relax.

I was so nervous and worried. I mean, I spent so many years sort of hustling, and worrying about what's next, and you know the whole nature of creative endeavours and freelance life means that you're sort of worried about what's coming next, and would I be able to eat and survive and so on. So I really think that persistence—and an appetite for knowledge, and for getting better and doing good work—means that you get to keep doing your work.

So I didn't know how any of this was going to work out at the get-go. You will get better, and this isn't brain surgery; nobody's going to die from a bad EQ!

Dave Hill

Dave Hill is a technologist who has designed many of the tools that audio mastering engineers now swear by. Most of his biographical details are available in this interview.

Q: What is your background and how did you get into making high-end audio products, many specifically for mastering?

Well, I don't know if any of it was originally specifically for mastering or not, but it kind of evolved that way. I had done studio stuff, and then of course live sound before that in the early '70s. At one point I met a guy who was basically looking to buy used tape machine stuff, which then became the Summit Audio products, and I designed all of that, and then we found people were saying, "Oh, this is really transparent," and I'm thinking, "It isn't transparent, it's actually very coloured," but you tend to learn a little bit as to people's conceptions or misconceptions of what they think things sound like, or how they describe things, because there are some issues with all of that, particularly on the side of compressors. Then, of course, there's the modern trend of bringing back things that probably should have never been used in the first place, but people are using them and [using them for] mastering or [on the] bus or otherwise, and it's like, "Dudes, that was an ugly sound back in 1975, why do you want to deal with it now?"

Q: Why do you think that is?

Well, I think there's some nostalgia thing, and this is a little bit not in the right direction, but it's kind of a good example. I went out for pizza with my wife recently, and they have some speakers in the ceiling and back-ground music going, and it's like, "Okay, this is Stevie Nick's last thing that she did," and that kind of vintage stuff they're playing. I'm listening to it and going, "Either the woofer is burned out in this, or the amplifier's got severe issues, or something's going on." This stuff sounded like thin and edgy, and just really horrible actually with high levels of distortion. Then a Beach Boy's tune comes on and it sounds full and lush, and it's like, "Uh-oh." I think there's been such a trend to make things a bit louder, "Well, gee, it's got to be louder," or "Can you tune?" and this is a question that actually some studio mastering guy came up with; there was some room somewhere in the States where they wanted the acoustic guys to make the speaker system sound like iPod mp3s, for doing mastering on, and I'm thinking, "Oh, this is where it's going to be." Well, I've always kind of believed in, you know, "Make it as good as you can. If you don't make it as good as you can, then what are you serving?" You're not doing your studio client any good, which gets into this whole "volume wars" thing. None of that stuff really makes it sound good, and when you get into it, of course louder always wins to an uneducated ear, but if you either go back and match levels or you have an educated ear, you're going to go, Well, yeah, it's louder, but it's ugly." As much as in music, some of it is the space, it's the space also that works, and if you just make things really dense, then all you do is become square waves at some point.

So in designing stuff, the first Crane Song product was STC8, and it really took almost two years to develop, because I tried different things, and it was going to be noisier. There were a lot of issues, and I wanted something that sounded different than the standard that was out there for compressors, hence the pulse-width modulator thing, and doing it discrete is what I ended up on. Even [during] the very first listens, when the thing was just starting to work, [it] was like, "Actually, this is quite good," and it got developed further into what became the STC8. There was a thing going on at the time where I was recording an a cappella of four women, and STC8 has the peak limiter in it that was designed for peaks, not for making things loud, like so many plugins now, which probably has done more damage to recorded music than anything [else], and if you've done recording, and you're doing an a cappella thing, and all of a sudden there are handclaps, you know you're going to have serious level issues. This was live to two-track tape at the time, and the peak limiter and everything, and the compressor kicked in a lot harder on STC8, and it's like, "Well, it's not distorted, it's really not bad sounding. Yeah, it's been squashed just to deal with the handclaps, but it basically worked." There were not bad artefacts other than the fact that, yeah, you could hear that you really whacked the amplitude of the handclaps down from what they were. "Okay, so this gives you some real peak kind of control," which at the time, and this was in the early days when converters were really starting to exist in the big push into the digital world, which actually I tried resisting for a while, but me resisting it makes no difference at all, but I did jump into it and bought a bunch of stuff that basically sounded bad. I really kind of waited until things actually started sounding a little bit better than they did originally. What happened is, of course, that the STC8 became a hit right away; mastering guys loved it. They loved it because there was really high accuracy and tracking between the channels, and it was very transparent and very musical at the same time.

Q: Do you mean like it sounds clinical almost?

Yeah, and there is a place for that, particularly when you're doing mastering things, but that wasn't really what STC8 was about; it was about having enough musical character to it but giving you lots of control and having that control to remain transparent. Whereas the tightening compressors that I do can be, well, clinical is not the right word, they can just be really transparent, but they do have a lot of colouring things in it so you can make it do "non-compressor" things. That's a little bit of a different kind of thing all around, but the STC8, it kind of surprised me that it really ended up in as many mastering rooms as it did.

The real next thing that I was working on was the HEDD box, which is now like HEDD-192, but the very first version just did the tube emulation, not the tape emulation, and it only went to 48 k. Early on, the converters just weren't that good at that time, and the components just weren't that good either. Some of it is how you control jitter, which was something that was vaguely talked about somewhere, but nobody really understood or believed it. Jitter's sort of an issue now. There are studies that say this is all you need, but that's really not exactly accurate and that clean of a statement, so the thing with the HEDD box, too, it was to make it so that it remained as musical as possible, and part of that was because of the colouring stuff, and part of it was just that I happened to pick good converters. It was always such that, as things evolved with the technology, I quickly learned that if you're going to build a digital box with converters in it and you want some longevity, you make it so that you can upgrade the critical sections of it—make the converters upgradeable, which the first version did not, but once we did the HEDD-192, the converters became upgradeable, and it's at the third or fourth version, and it's about to get another version of them before too terribly long here, and that's really to (A), take advantage of improved sonics, and then [B,] give people something that, "Oh gee, this is two years old, and I've got to completely throw it out now, because there are new converter parts." I'm a small studio here and face the same thing that everybody else faces, that you don't have unlimited money to go out and buy new gear every year, even though there are manufacturers and stuff who would prefer that you did that. You know the saying—sort of like a "10% growth per year forever"—well, that doesn't work, people. You get to a point where everyone has to buy a new one every month in order to sustain that! So this upgradeability becomes an important thing.

So it kind of all evolved, where I wasn't necessarily thinking about mastering, but it did evolve into that just because the stuff worked really well. That's part of when I design gear, it's like making a mix; you design the thing, it works. Then you go listen to it, and then you tweak it, and go back and forth. It's not like the people who just go, "Okay, well, the spice simulation says it sounds good, so we'll go with it." That's just not how I've worked with things.

Then the other thing that kind of happened is, when I went to do an EQ. At that point it was the IBIS EQ. I said, "Okay, so let's make this so it can be really transparent, so it can have a very short audio path but be smooth at the same time." So from a design point of view, it's an input amp, an EQ amp which is a single amplifier, and the output. So a shorter audio path. In other words, the less crap in the audio path, the better things sound, is the reality of it, and it's true in digital design, and it's true analog-wise, too. The less junk you've got going on, usually the better off you are. So when I designed the IBIS EQ, first of all I looked for a filter that really wasn't being used, because I didn't want to do something that was like what everybody else was doing. I also went, "Okay, well, being really clean is good for a lot of people, but a lot of people are going to want some colour." So I figured out a way to have some kind of adjustable colour in it, and then we made it so the EQ could be built with switches in half-dB steps. So you could actually do a mastering thing with it. Once again, that's been an experience; if STC did not find its use in mastering, and if HEDD did not find its use in mastering, then probably I just would have never built it to be able to do that, but it just sort of happened.

Q: We're fascinated by what I'd call the paradigm you follow when you're creating these things. The Avocet in particular is in so many mastering studios; would you take us through the design and innovation process for that, from an inspiration level?

The Avocet is kind of an interesting story because it almost happened to be the right thing at the right time, too. A very original version was basically a two-rack space [unit] with a gain knob on it, a few switches for input selection, and a D/A, which is nothing like what it became. I probably had six or nine months of toying around with the thing and getting it to that stage. I was using switches for step attenuation, and it's a strange thing that way, and it's really not the first monitoring device I'd ever built, because I had built other stuff for my room prior to that. I never intended it to be a product or anything like that, but had some basic switching and a gain pot, some buffering amplifiers, and that sort of thing. A friend said, "Well, why don't you make this thing so you could have it remote controlled?" I believe I did have it at an AES in New York, just the fixed unit with no remote; the very original version of the controller, and so nine months later it's became remote controllable.

Part of it was originally that, "Okay, I'm a small studio, I need an accurate monitor section." I had an old Spectrasonics console at the time, yet you turned the gain up or down and the image would shift, it was like, "Arghh!" You can't function that way, and I've always believed that if you can't hear what you're doing, how do you know? The example I would use for non-audio people is, "You're going to put on your tinted glasses and go paint your house. Select the paint and paint your house, and then take the glasses off and you wonder, 'Whoa, what's wrong?'" Monitoring is basically the same issue. So as Avocet developed, essentially I wanted to keep it as intuitive and as purest as I could. I didn't want it to get in the way, to colour things in any unhappy manner; hence, the main audio path came as all discrete Class A amplifiers, and then the step attenuator, it's resistors and relays. Though a lot of it originally was 1 dB steps, and by doing that, it's mechanically a little awkward, but you've still got a couple of microprocessors, then 16,000 codes of assembler later, it takes a lot of awkwardness out of it, and it became fairly intuitive. What would happen is I'd be just going for my hike at lunch time and going, "Well, gee, if we could have a way when we select the input to trim the levels on the fly from one another for level matching." It was things like that where I'd go out and would be thinking about it, and go, "Oh, this would make things better." I have a friend called Brad Blackwood who's a mastering guy, and I would talk to him about it, and he would have some ideas and I would think about it, and it sort of just evolved, and it was really kind of funny because you build the thing, you get it out, then you hear things; it has always continued to evolve, whereas if you've got an existing design and you keep adding stuff to it, there's a point where it's like, "Okay, let's just start the whole circuit board all over again, put in everything we already know plus some new things." Then you do that, and a year later it's like, "Well, can you add this?" Well, it's an analog box, and part of my philosophy on this is if you're doing monitoring, mastering, or otherwise, yes, there is some sort of advantages to all of these speaker systems with the digital crossover and everything, but what about those converters in those speaker systems and the jitter in those converters, and the bad filtering in those converters, so all of a sudden you have a speaker system that is permanently coloured.

So as Avocet evolved, I've always continued to look for ways of getting it better, and it's the fifth version of converter in there right now on the D/A side, and I've always taken the sample rate converter with a reference oscillator to try and reduce jitter. The whole filtering and sample rate thing is really extremely dumb when you think about it, because in designing servo systems for motors, say for elevators or for your underground system, which I'm not sure what's new in these days, but anything that is designed now, the engineers go, "Well, if I want a 20 kHz bandwidth on controlling this motor, my sample rate is going to be 10 times that, and that's what the servos do." But with audio, "Oh, we'll just do it two times that, so we can cram more of it on this disc."

It was really a technology that probably was not ready, that's the reality of it. So what happened is that new converter parts would become available, and you build up the thing and you do an A/B, and a lot of companies [would say], "Oh, well, we'll send you the evaluation board." [I'd then say,] "Well, that doesn't work. I need to know how the part works in the system." So you'd go through, and you'd build it, and you'd do a careful level match and go, "Is it better or not?" Sometimes it was dramatically better, and when I started going down the clocking thing, which right now I'm measuring a new batch of crystals that I just got, and I'm somewhere in the 19 out of 20 of them, I'm somewhere in the 0.3 picosecond range of jitter measured from 1 Hz to 100 kHz, which is crazy low because the logic circuits are worse than that. What would happen going down that path is you would build a new D/A, you would put this, and this took six to nine months to be able to build a clock to make this thing work, and you would go do a careful level match, and the first thing was, the older one sounds better, what's going on here? Then you would listen, and this is the funny thing that happens with the [whole] jitter thing; jitters can make things sound warm to an untrained ear. Well, really what it does is it makes it muddy, and if you spend a half-hour doing some A/B [testing] and you listen to different material, you think it's not warmer; it's just mud!

So now the filters are a little bit better, and in the last round of it, what's ended up happening is the filters [were improved enough] that the transient response improved greatly, and this low-midrange thing that happens, the smeary stuff, is mostly gone, and [then you'd] just see what happened, so it was just more and more accurate. When I played the prototype for my tech support guy, the first thing he said was, "It sounds like I'm in the room with the drums." The transients work so much better, but the funny thing with the whole jitter thing, and there's papers out there that say before a certain level it doesn't matter, but crystal oscillators, they have what they call "phase noise," and that's noise from the centre frequency, whether you're 1 Hz out or 10 Hz out or 100 Hz out. The crystal oscillator wobbles in frequency in its phase noise, as they call it.

Some of this, I think, is the result of choices in converter technologies, as they tend to be more sensitive to jitter than other particular designs, but with anything, it's a trade-off; if you design a circuit that solves one problem, well, what are the other new sensitivities you have to worry about, and how do you solve those particular problems? So the latest converter, which for marketing we're calling it "Quantum D/A," basically there are typically less than a half-picosecond jitter from 1 Hz to 100 kHz. If you measure it from 10 Hz to 20 kHz, or like .05 picoseconds typical, which is ridiculously low, and on papers that claim what happens at 10 Hz and below doesn't matter. The hi-fi world believes it matters. You can hear it, because I've done listening tests, that phase noise of 105 at 10 Hz versus 110 or 115, and done careful A/B listening tests; the more jittery the −105, it's muddier sounding, the transients aren't as good, you lose clarity, so it is definitely an audible thing.

**Q: How long do you think we've got to go before the design of convert-
ers are up to where you think they should be?**

I think we're just getting to the point now where we have materials to
work with and [enough] understanding where we can actually do good
conversion, and just getting there. It turns out that in the circuit, everything
matters. There's so much detail you have to pay attention to, but also what
happens is there's this hearing education issue that has to happen, because
these people hear the old muddy thing and think it's better, then you end
up fighting that. There's a bit of an education issue that takes place that
you do have to deal with. How long it's going to take? Here's part of the
problem with the "how long" question: and at what cost?

So first of all, the rubidium oscillator is more bullshit than reality, and
the reason for that is, rubidium's really good if you care about being within
picoseconds next weekend; it doesn't do audio much good. What we care
about is what happens in 10 ms or 100 ms, short-term kind of stuff [which]
is a lot more important for audio. If your references change by a nanosec-
ond from one day to the next, but you're still at a half-picosecond from
one minute to the next, well, that 1ns drifts so slowly that it doesn't really
matter. It's kind of like, "Well, then, why did it change?" The thing about
quartz is that things like gravity and all sorts of things start affecting it.
Where the moon sits can actually affect it a little bit; it's really quite sensitive. There has been a lot of research done towards doing really low jitter
clocks and building really good crystal oscillators, and this has been the
result of microwave communication and space communication, because if
you start at 10 MHz, and you're going to multiply that to 20 GHz, well,
any variation also gets multiplied.

So getting back to it a little bit, there's a reference oscillator for building
an A/D/D/A that a company called MDK supplied. For a quantity of one
they're $1,500, so if you're going to build a D/A you might get away with
one, maybe you'll need two of them. If you're building an A/D, you'll
definitely need two of them, so that might give you an oscillator that is
pretty much as good as anything that's humanly possible at this point.
It will help clean things up, but this gets to be a little bit like, "at what
cost?" Even if we could say that this clock is perfect and cannot be any
better in any form that's audible, we're still dealing with filters, and we're
still dealing with time domain problems. That aspect of it is that at 44.1,
you're essentially screwed, just because of the requirement of the steep
filters. They do a compromise now, it's called "minimum phase." It's part
linear phase and part non-linear phase, so the pre/post echo time domain
issue, transient response issues basically, they become not as bad as if it
was all just linear phase. You get less of this pre-echo. You'd have more
time-smear with just the linear phase only. So, the combination of linear
and non-linear gives you less time-smear; unfortunately, if you did the
filter so there was no linear phase and no pre-echo, you could get to the
point where there's so much phase shift in the audio that things sound
bad anyway, so it ends up walking a fine line. However, as you go up in
sample rates, the higher the sample rates, the less this becomes a problem;

44.1 to 48 k, I know people who claim it makes a difference, but I don't think it makes any real difference. You get to 192, and you're in a different world because the filters aren't as steep; hence, you don't have as much pre/post echo or time domain issues, so the filters become a little bit less of a problem because you need less filtering. And if you have really good clocks, all of a sudden things start sounding really a lot better. You get to a point where you can record something and play it back, and it sounds the same. If you go out and listen to it in the analog domain and then listen to it digitally, things are getting fairly good. Now, unfortunately, there's a part of the market that's saying, "Oh, it's going to be an mp3, so it doesn't matter." Well, that's not true, because the better it is to start with, the better the end result is going to be, no matter what happens in the meantime.

Q: So we're only just now coming of age in real understandings of jitter and sample rates?

It's all a little bit, kind of, complex. In analog, you can have better resolution and bandwidth any day of the week, as it's easy. If you want a 1 MHz bandwidth, with an analog pre-amp or something, it's really easy to do. A friend of mine did a little bit of a recording school test, where he set up and did a recording all analog to a 24-track with a bunch of students, and then the students did a digital comparison. The students went, "Oh, the analog recording coming off the tape sounded really good, a lot better." Then he asked them, "What did it sound like before going onto tape?" They responded, "Oh, that sounded really good, too." It wasn't the tape, it was the fact it was all analog, and at least the students got it, but I think we really need to be working at 192 minimally, and it makes a real difference in a lot of things. Some of it's just weird distortions that you can't explain, and once you go to 192 a lot of these problems just seem to go away. There's kind of another aspect of it because 24 bit is pretty much the deal. There is some push and talk about, and some of the converter parts will do 32 bit. Well, I'm not sure that there's any real audible improvements with it. However, one of the things that can happen is, like the convertor chipset that I'm using, the sample rate converter outputs 32 bit, and the DAC will take 32 bit. Before you would end up truncating or dithering the output of the sample rate converter before it went into the DAC. Well, maybe you would get an improvement because you don't have to do that, maybe not. I think this one is a little out at the moment, but I haven't done any real listening tests to go one way or the other, because it's a little bit of a tricky thing, and the reality is you're not getting a lower noise floor from the 32 bit.

Q: You're getting into the realms of you could dither digitally from 32 bits, and you'd end up in a different place.

Well, I use a silly example; even at 24 bit, which is like 144 dB dynamic range, so even the good converters were maybe hitting at 19, 20 bit noise floor, we're probably not getting much better than that. I think there is some stuff claiming a 126 dB dynamic range out of it. How do they measure that? It's in the noise floor anyway. So how do you get rid of noise?

Well, cryogenic cooling, which they use for radio telescopes, right, so let's see, if we really need to be that quiet, we could cryogenically cool the microphone, the room, and the singer, to get the noise floor lower.

Q: You start to get into the realms of, if analog is the de facto standard that we're trying to reach, and I know that sounds like a backwards statement, but that's what everyone in the vinyl revolution and so on is currently saying—if we're trying to exceed that, we're asking ourselves these questions.

Right, definitely. Analog still does seem to be the standard that we are trying to reach. At 44.1, well, even 96 k, you're not going to make a plugin that exactly emulates an analog device. The time domain issues that exist in the digital world kill it before you start. You can still make things that sound good, but it's not analogous. Massenberg did a TDM compressor, and he actually gave me a license for. It sounds really, really good, and he's put years and years into it. However, the analog box still sounds better. I did an A/B comparison on it, and it's like, "Yeah, this sounds good, it's very useful," but you pull out the analog box and it sounded better. There's something that happens. So when I did Phoenix, the tape emulation Plugin, it was TDM fixed-point 24 bit, and that's the kind of programming world I live in, because I taught that kind of stuff back in 1975 or '76. The thing about 32 bit float which is interesting is when you're doing things with filters, whoever is coding no longer has to pay attention to dynamic range issues or where clipping is. It makes it easier to code stuff. Hence, you could say, "Well, maybe the 32 bit sounds better?" And you think, "Well, maybe, maybe not. If the 24 bit was coded properly, the 32 bit might not really have an advantage." So it's not as straightforward a thing as everybody who wants to sell you 32 bit float makes it out to be. It's easier for the programmer to get there, because there are a lot of things you don't have to pay attention to.

Q: What's your view on floating points in conversion, because it's often described for those in Pro Tools as having almost more headroom in the system, whereas those in the music production end of things using Logic are kind of hamstrung by their audio engine?

Well here's the deal: the converter is 6 point in the end anyway. Stage-Tech has made some converters for years and years that are sort of a quasi-floating point kind of thing, so you didn't need mic pre-amps and stuff. It's a theatre kind of deal. I know they were really decent sounding converters, and I haven't followed that for years and years, so I don't know where it's at anymore. AES/EBU or SPDIF, or your main data path are all 24 bit, anyway, and so this chipset I'm using, well, it's 32 bit 6 point; it's not floating point going into the converter. Going into the sample rate converter, it will take 32 bit in, but AES/EBU use 24 bit anyway, and the extra data bits, I don't think give you much of a sonic improvement. Going back to Avocet, it has a button on there that truncates the incoming data to 16 bit, and this was something that Bob Katz suggested, and then when he wrote the review was like, "Well why the hell did you do that?" I responded by noting, "Bob, it was your idea!" And he went: "Oops." What happens is, if you have properly dithered audio, in a really good speaker system you hear a little bit of an image collapse and stuff, if you have improperly dithered audio, or [are] doing some weird tests, because we were doing a low/high resolution test, how well of a signal could you hear, kind of thing. Well, truncating to 16 bit really makes a big difference, it can make things really ugly, but if you have a really bad speaker system, you may not hear it anyway. So, resolution in speaker systems is also a part of the whole equation.

Q: Tell us more about your development of plugins.

Well, the history with the plugin is that I had the HEDD box out, and the Digidesign guys said, "Well, can you make a plugin?" First of all, I have issues with high level C programming because I'm an assembly language guy, and then this was at the time when they were going from OS 9 to OS 10 and that transition on a Mac—what a nightmare that was. I met a guy and brought him in, and he did the C stuff. I worked on the algorithm side of things, and we did TDM only, and of course as soon as you start getting things to work, "Oh by the way . . . so that's no longer going to work," so you have to redo everything, and all of this and all that particular nonsense. What I've done is, like with Phoenix, is it exactly tape? No. Can you ever make it exactly tape? Well, how much computing power do you want to dedicate, and will it ever really do that anyway? Well, first of all you'd have to undo the time domain issues, which you really can't, it's the physics of Nyquist basically, but what I do is I make it so it's a useful sound. I still record, I came out of two-inch 24-track land, so I still record enough that it's got to be a useful sound. You do what all the theory says to get there, and then you screw with it until it sounds right. So there's like a waveform compression thing, and there's a vinyl emulation thing, which is also kind of in the ballpark. Is it exactly perfect? No, but it gets you kind of that sound, and then of course what became Heat, which is part of Pro Tools. I'm dabbling with some other plugin things also. I've had people with other systems say, "Oh, you'll let us take STC8 or IBIS EQ and implement it as a plugin on our system?" I'm like, "Uh, no, because it's not going to sound right."

Q: So, it doesn't sound right. Is there anything else that you would add to that, to say that you're not very happy with it?

Well, yeah. I think there's certain things to this day and age that are done better analog-wise and certain things you can do better digitally, so really, take advantage of both. The one thing I like to use as a comparison with this whole emulation thing is, I just say, "Well, weather forecasting is a modelling emulation thing, isn't it? That doesn't work particularly well either."

Q: What do you see as the future of mastering from your perspective?

Well, there's a couple of things that need to happen. There are efforts to, and Apple went down this path with MfiT [Mastered for iTunes]: "Okay, this is what we want to see for our masters," and yes, it does sound better by not turning things into square waves and that sort of thing. I talked to some other friends and mastering guys, and they basically said that right now things are a bit of a nightmare, because half of everybody follows that mindset, and half of everybody doesn't follow those guidelines. As a result, it gets to be a real problem. I think, however, mastering engineers need to be allowed to go back and make things sound good. If you're going to have some sort of target RMS level, fine, but don't take the peaks out of it, try to put life back into it. Put the dynamics back into it. It goes back to this thing that I heard in the restaurant, and my first thought was, "Have things really deteriorated to that point where things really sound that crappy?" And then you think, "Well, what is some of the resurgence of some of this old gear, and the need to emulate it all? Are we looking for something that we lost, because what we're doing is basically flawed?"

You know, it's kind of a valid question, because we're destroying things by making things loud, and then we're going back and looking to have this sound that we no longer have that's been lost. Isn't the process of destroying things with peak limiting and making it loud kind of undoing what we are attempting to do in the first place?

Q: We went to interview Ray Staff while he was doing some of the vinyl cuts of the David Bowie remasters, and obviously listening through your Avocet. We were just talking about some of these things and how he's approached that in a remastering setting, and what freedom he's been given. You're absolutely right: being allowed to go back to the master tapes, transferring them across, and not having to muck about with them in terms of making them "brittle" or "loud" and so on.

Everyone of that experienced generation are saying that we do need to return to something that does have that RMS level, and it's almost supporting the EBU R128 standard and other LUFS based standards that have come out since, and that's a really, truly hopeful place that we're going to end up in. It's great to hear you say that, especially from a manufacturing point of view.

Ray actually got one of the early prototypes of a Quantum D/A in his Avocet, too, so there was some feedback from him which proved helpful, and found some things that kind of misbehaved occasionally.

I had another thought, because I did a Christmas thing the other night for a guy from a radio station, and we had to pull tunes off of his iPhone, which turned out to be a real nightmare. I ended up plugging a phone jack into the thing and doing it analog. We couldn't get the damn thing to work and talk to the Mac properly at the time. Apple on the surface is going, "Yeah, we want it to sound better," and blah blah blah. But there's another thought that occurred to me, which is that these devices are so low power, leaving more dynamic range in the recordings means it consumes less battery, and I just wonder if that was an influence?

Q: That's an interesting point. It's interesting you say that, because no doubt you've followed the whole Pono thing from Neil Young, and one of the reasons why it was made so triangular and so large is because of

the power concerns from the pre-amps and so on, so I think there's a lot of chime in with that, to be honest.

There are so many debates to have, and I'm not a cynic, and I'm not of that nature, but It's almost like we've been fooled over these years to think digital is fantastic, but we've lost the true essence of what we're really after. Well, and the other part of the whole recording thing is when people don't play together, when things are just pieced together, that you lose some of the magic, too. Many years ago, I did a jazz fusion thing; some friends and Ernie Watts came to town here to play on it. The saxophonist and Ernie had played with Johnny Carson's band, and I think he did a tour with The Stones and is an incredible player. This was a two-inch 24-track thing, and we did overdubs on three songs, and three live tunes with them, and with the live tunes, it's like, "Oh no, not even a rehearsal take? I just have to hit record and have it work? . . . Okay." So not knowing about levels and stuff, well, it turned out fine in the long run. However, the big question to Ernie was, "Well, what do you prefer?" He said, "Well, playing live; it's always better."

Bryan Martin

Bryan Martin is a founder and CTO of Audible Reality. Audible Reality creates audio technology for virtual and augmented reality. His obsession with audio began in 1977 as a twitchy 16-year-old thrust behind the console of a touring rock band. From 1986 to 2002, his time was split between New York City and London as a record producer and engineer, where he worked with such artists as David Byrne, Rufus Wainwright, Run DMC, Max Roach, and the Cars.

In 2002, Mr. Martin moved to Montreal and opened Sonosphere Mastering. While there, he worked with The Pretenders, Pat Metheny, and Zappa Plays Zappa and garnered Grammy, Juno, and ADISQ awards. As a PhD in the sound recording department at McGill, Mr. Martin pioneered methods and techniques in three-dimensional music recording. He lectures at conferences and gives master classes on the topics of 3-D audio, microphone arrays for vertical imaging, mixing in 22.2, the design and construction of vacuum tube audio circuits, classic guitar amplifiers, music mastering, production, and recording.

Q: In your words, what is mastering?

Mastering is the optimization of the transfer of information between the artist and the listener.

Q: So how did you become a mastering engineer?

I became a mastering engineer by default. I was a record producer in New York, and I used to bounce back and forth between New York and London quite a bit. I had an apartment in New York and London, then in the mid-'90s, I actually became a partner at a studio in Williamsburg called Excello, and that's when the record budgets were going down, so I wasn't travelling as much. Then around 2000, the record industry was really declining, and I was doing more Pro Tools manipulation than actually recording music, and I was getting sick of it. My wife's from Montreal, so she asked me if I wanted to move, and I agreed because it wasn't as fun as it used to be. So I came to Montreal. I had no idea what I was going to do, and a bunch of people suggested that I should open [up] a mastering studio, which was the stupidest thing I ever did; however, I spent a year building it, and it was great. There was a lack of mastering here, and that's why I became a mastering engineer.

One thing I found out was that the way you think about mastering music has nothing to do with how you think about recording, producing, or mixing music. It's a completely different way of thinking about music. You use EQ differently, you use compression differently. I think it's really difficult to bounce back and forth between being a mastering engineer and a recording engineer. A lot of people say, "I mixed it, and then I mastered it," and I think, "How could you do that?" Firstly, in the mix process, didn't you do the best you could? So in mastering, then how can you have a fresh perspective? In these cases, the mixing and mastering is usually done in the same studio, so if there are problems in that studio (and every studio has idiosyncrasies and issues), then you basically have no objective view on the music in the final mastering stage. And you've compounded any problems by trying to master it in the same place it was mixed. When I've had to remaster those records, the mixes are generally better than the master because when the mix engineer went to master it, he already did his best and then just tried to do something else, and that usually degrades it.

In my view, you need that objective person who's unattached to the record to look at it, thinking, "I'm going to optimize this music, I'm going to optimize this message," rather than somebody who would think, "I've already done my best work, and now I'm going to try something else because I'm supposed to, because that's mastering!" So that's how I got into mastering, which was kind of by accident.

Q: How do you think mastering is different to mixing?

Mastering is completely different from mixing and recording, because you have to think about the music in a completely different way. You need to think about the song as a whole, complete message, the CD as a single artistic statement, not a sum of separate individual instrumental parts. You don't use any of the tools (EQ and compression) the same way. It is not anything like the mental processes that go on when you're recording or mixing. It's a different process completely, and I think a lot of people don't know that, except for the mastering engineers!

Q: So how did you learn to master, then? How did you first sort of catch on to these differences?

Well, I made records for a couple of decades. So I *thought* I knew what I was doing, but when I tried to do the same things that I used to do in the recording process in mastering, it didn't work. So failure is the best way to learn, and I was failing. Because I think a lot of times when you're a mix engineer and you go into mastering, you view mastering as a remix process, and I think a lot of artists view it as an extension of mixing, which it is not. With mastering, you have to look at the whole, and everything you do affects the whole, so you have to look at the music as one unit, not little pieces, not just separate elements of the song, but as a complete song, and you have to try to optimize the music: the transfer of information between the artist and the listener.

A lot of times, there's a lot of damage that's done, especially with [digital]. Digital tools are incredibly powerful, you can do a lot of damage, and you have people who are uneducated, who are self-taught, and maybe have a lot of bad habits, and on top of that the quest for volume. All that can inflict a lot of damage on the music. So if you have bad and excessive processing, plus excessive volume, a lot of damage can be done before the mastering engineer gets it, and I think a lot of us would say that we do a lot of repair work in our efforts to get to the actual message of the music.

So everything is experience, and I had the albums I produced mastered by a lot of very talented guys. I worked with Greg Calbi a lot, and Scott Hull, and Chris Gehringer, and Howie Weinberg. In London, I used to go to The Exchange and Metropolis. When I was working at The Townhouse I'd go over and talk to the guys at Metropolis, so I have a lot of friends who are really great mastering engineers. I attended around 100 mastering sessions, so I wasn't completely unaware of what the process was; it was mainly just the mental switch.

**Q: Either when you were observing or when you were working your-
self, are there any mastering sessions in particular that stick out as
being instructive?**

I think in the analog days, when the processes were more tactile, more hands
on, as opposed to in a DAW, I think those were more instructive because
you saw what was happening more readily. I also think that you cannot get
educated about mastering until you learn how to listen, and to learn how to
listen you have to spend thousands of hours working with music and music
people much smarter than yourself. The first couple of times I went to mastering as a young engineer, I didn't know what was going on. It was just this
crazy, overwhelming, wild display of technology; I had no idea. I think over
the years (and I mean a couple of decades of going to mastering [sessions]),
I never ever told any mastering engineers what to do. I would just go in
and pay attention and really listen to what they were doing. Sometimes I
would look as they twisted the knobs, but most of the time, it was about the
education of listening. Until you can hear what's going on, you can't learn
anything. With the kids that I teach now, I play stuff, and they can't hear it,
and until they can hear it, you can't teach them. But once you teach them
how to hear, they really start to zone in; but that's a process.

**Q: When you're mastering, what are the things you typically listen for
regardless of genre?**

Western music is made up of three main elements: there's the dynamics,
the harmonic, and then there's rhythm. Those are the three things that are
really the concepts of music. So when I listen to a track, I'm listening for a
degradation of any of these elements. Usually, the first thing I listen for is,
"Is the groove working?" For me, if the groove isn't working, the listener is
never going to get it, never going to feel it, to internalize it. Where it has to
get you is in the hips, not in the head, below the belt. So when I listen to pop
music, I'm thinking, "Hey, am I grooving? Is it getting me below the belt?"
And if that groove isn't working, then I have to look at why it's not working and try and sort that out. That's one of the things I always look for, and
it usually has to do with overcompression, the destruction of dynamics. So
that's a problem, which means you've damaged the rhythm, and if it's super
hot level-wise, the distortion damages the harmonic content. So I try to
minimize this damage and bring back some dynamic and harmonic clarity.
   And then, of course, always in pop, you ask, "Is the vocal clear?"
Because if the vocal's having trouble fighting through the track as well,
that's another problem. So really the things to focus on are the groove,
is the vocal clear, and then the distortion element, and if you optimize those
three, then the song has a chance to translate pretty well.

Q: Do you have a preferred chain of gear that you work through?

I think if you walk into a mastering studio and see a ton of gear in the rack,
run for your life. Because think of it like those are a bunch of penises, and
all you're doing is masturbating. All the engineers I've worked with and

respect have a chain that is never touched, because if you have your chain, you know what it's doing. It's basically your reference: you know what it's going to do and how it's going to react, and if there's something wrong, it's not you, it's what's coming in. If you're jumping box to box all the time and trying this and that, my question is, "Why don't you know what you're looking for?" Bob Ludwig is the only person I know who can listen to a track, and he'll know, "Oh, this piece of gear will make it right." But I mean, he's God, and there's only one God. So I think most of my buddies, who are the old guys, all have a chain, and we never touch it, and I have one, too. Basically I have a Z Systems EQ and I have my Weiss mk2 compressor and that's it, and I never ever swap out. Well, once in a while, there might be some crazy, mangled, destroyed, heinous, sonic disaster that needs something special, but 99% of the time, I never touch my chain; it's a constant, and I know what it's doing, and I know how it's going to work, and I never deviate.

Q: What about converters and monitors?

I have Prism Dream converters; they're great. If you have a really high quality converter, and there's a few of them, you're going to be fine. I think that can be personal taste, but many are really good now.

I think a lot of times the monitors in a mastering session are basically for show: the bigger and louder, it looks great, but do you really need the insane monitors? I would say no. For monitors, I think you need a really good wrench. Mine are actually kind of small; I have the Dynaudio Special 25s, which Dynaudio made for one year to basically say, "Look how great we are," and they're great. They're a two-way speaker, but they go down to 30 Hz, and they're super clear. My theory is that the more drivers you have in a monitor, then the more crossovers, more crossover distortion points, and it's just a more complicated system, so you're going to have more anomalies. I use the Special 25s, and I love Bryston amplifiers, so I think simpler is better, in mastering especially. The less tools, the simpler the tools, the more purer the signal path, and the less damage you're going to do.

Q: What about a workstation?

I use two. I use Pro Tools HD for some things, and Sonic Solutions, depending on what's going on. Sometimes people send stems, which I hate, so I'd prefer to do that in Pro Tools, but Sonic Solutions, I've been using them for a long time, way back when it was quite cumbersome, but those are the two I use.

Q: Can we talk quickly about stems, about what you don't like about them?

Stems are a bad idea made up by mastering engineers who want to be mix engineers. That's the first part of it. "Why are you sending me stems? Whose record is this? It's not my record? That's a mix process." So any mastering engineer who wants stems, I feel, is trying to mix the record: that's on an artistic level.

Now on a technical level, when you send stems out of a workstation, if you had a bus compressor on the whole mix, which they all do, it reacts differently to the lesser stem content than on full mix content, and the stem content generally falls apart. For example, the reverbs won't work the same, because you're changing what's happening on a whole lot of levels, so the stem mixes can't be the same as when the full mix content is feeding them, i.e., the thresholds and so on. I really think that the whole stem thing came about with people who really couldn't master and had to look like they did something. I think it's an ego control thing; it has nothing to do with art. If the producer and the mixer want to make the record, then have the balls to man up and make the record. Commit; don't hand off the responsibility to someone else in hopes they will save you.

Q: So let's talk a little bit about the relationship between the mastering engineer and the mix engineer there in general, because it is interesting.

Well, I think that, and I'm a cranky old man, the current quality of mixes have been degraded for a few reasons. Firstly, the quest for loudness has destroyed the quality of the mixes. Secondly, the studios are gone, so people are working in home studios, usually under less than optimal monitoring situations, and maybe even less than optimum hardware situations. I also think that the quality of engineering is degrading, because with the studios gone we lost the universities of recording, our great accumulated centres of knowledge. We don't have the transfer of information, and we don't have the apprenticeships where the assistants soaked up the decades-long accumulation of experience. Now, people buy a workstation, and they hang out a shingle and claim to be an engineer. The chain of the passage of knowledge is broken, and the current generation does not have the opportunity to record in great facilities with great teachers. It is broken.

What we are left with are uninformed people making mixes, and on top of that we have the artist demanding the volume thing. The artists, I'm pretty sure, drove part of this whole volume thing, so there's a lot of factors contributing to the destruction of fidelity. So the first thing the mastering engineer has to do is try to minimize the damage and do some reparation before mastering.

Q: How often do you find yourself being leaned on to increase loudness?

Probably about 85%. It's basically, "make this loud, make it as loud as anything," that's a given. Over the last few years, it has actually declined a little bit. And even with loudness normalization in, I have not seen the artists or mixers change their loudness habits. I would say, and you can ask any mastering engineer, you can't send it back quieter than you got it, otherwise you'll have a problem with the label or the artist.

Q: Can you talk a little bit about equalization?

One thing that happens with the overuse of compression, or can be an arte-fact of bad math in plugins, is that you can get enharmonic distortions: frequencies that are not in (related to) the music. These frequencies may also exist at very low levels but are raised up via excessive compression and limiting to being equal with the regular harmonic content.

What generally happens is, there is a build-up of low frequency sludge below 40 Hz, and a lot of times you can also have a lot of noise generated above 10 k, which is also enharmonic because it's bad math spitting out noise frequencies. One Band-Aid is to create a very narrow notch filter. A lot of times you don't really hear either of these artefacts very well until you get rid of them, so if you build a narrow notch filter dipping 6 dB, you just start it however low you can and start sliding it up slowly. A lot of times you can't even hear this because your speakers aren't outputting these frequen-cies, but as soon as you hit the problem spot, all of a sudden your bottom end will pop out and become very clear. And the same thing at the top end: if you create a notch and start at 20k and slide it down, you will hit one of those noise spots, and all of sudden you'll feel less pressure on you as you cut out this noise. So both of these actually work quite often; everything else is specific, but those two problems occur almost all the time.

Q: Do you address stereo image at all, and if so, how?

I'll go on another rant. I think this whole widening thing is just another big masturbation event. Why do you have to get it wider? I mean here's the thing, how many dicks are in this mix, right? Everybody has to do something, and now it's this "width thing." Do they say, "I can't make it wide by good mix techniques, 'cause I don't know what I'm doing, so I'm going to let the mastering guy put this width thing on, and decorrelate my centre, which kills my groove and messes up my bass content, and also introduces a lot of really bad phase information." Then, of course, they want to cut the vinyl right, which makes most mixes today, with all the high and low frequency crap, impossible to cut. That's another thing that is really bad with the digital world. You can put anything on a CD, but now everybody's starting to cut to vinyl [again], and they give you these mixes that, in the vinyl age, would have been roundly rejected, because we have all of this out-of-phase and high frequency crap. So I think the width thing is just another artefact of complete ineptitude in mastering. I never do it; the only thing I do, once in a great while, is when it's so decorrelated, I will actually make it narrower so that the groove actually works.

Q: What are you using to do that? Is it some mid/side thing?

There are different plugins that do it. I mean, I can do it analog or I can do it digital, but digital does it really well. When I'm cutting vinyl, when I'm making vinyl masters, I always have to look at the low end and make it more mono, because otherwise it will not always cut out of phase; bass information makes the needle go up.

Q: Technically speaking, when mastering for vinyl, how are you preparing your vinyl masters that might be different, and also how are they aesthetically different?

Firstly, you have to look at the bottom end—it can't be out of phase. You can't have enharmonic sludge, and you can't have a lot of crazy junk below 50 Hz, so you have to always clean that up. You have to always make sure the bottom is coherent and correlated, because if it's not, it won't cut, so you have to deal with that. Secondly, you can't have all that trashy high end because it will melt the cutter head, and what a lot of the vinyl cutters do is they just rip off the top and the bottom and shove it in the groove. That's why current vinyl sounds horrible, because the people mixing for this have no idea how to mix for it, so that's always a problem. Thirdly, you don't always have to deliver the vinyl hot, +4 is fine, this is going to analog gear, so that makes it a lot easier for the cutting engineer if you deliver a master at the same level as something from 1985. So you really have to address those two issues; if your mix is hyper-compressed already, you don't have to even worry about the compressor, unless you're using it to enhance the music, rather than reduce the dynamic range.

Q: Do you have any experience remastering?

Yeah, that's always tricky. If it's a digital master that they want remastered, it's probably already so damaged, where it sounds so bad, so crushed and distorted, which is a disaster. You can't do anything if someone has already crushed it. I did have to do a few compilations where I had a slew of masters that went from vinyl to current era. That's really hard when you're trying to make all those songs sound like one record. That's really difficult. It's almost impossible, because you have very dynamic stuff from maybe 1975, and incredibly mangled stuff from 2005. That's really the hardest job you can get as a compilation, because the songs have nothing to do with each other, they don't sound like each other, so that's really a challenge, and it takes a long time.

Q: How do you approach that challenge?

Well, I think the first thing you have to do is to find out where the volume level is going to be: "What's the level of this record going to be?" Then you find the best mix that's going to serve the record as a whole. Then try to bring everything up to that level of fidelity, and also in the ballpark of the dynamics. So I usually pick the best [track] and then try to bring everything up to it.

Q: And then you're going to use that as a template for balancing and whatnot, right?

Yeah, you find the best track, and then you make everything else, as much as you can, get to that ballpark. Instead of using mediocre material and pulling the whole record down, you just say, "Look, if some of them are really bad, well it can be bad," but you want to try to get most of the record as good as it can be, and you need a target, and you usually find the best-sounding track, which is a good place to start.

Q: When you were mixing, were you asked to do different things by mastering engineers?

No, never.

Q: Do you ever mark back to a mix engineer?

You know what, unless the guy is really good, and there are a few of those, it's a waste of time. Most of the time when I would go back and say, "You have distortion blah blah, here and there. Could you deal with this or that?"—but they don't know where their problem is. Their mixes are so lost. They've done so much junk that they don't even know how to go back and find the problems. So, a lot of times, they get pissed off like you're somehow criticizing their self-proclaimed genius. So, unless it's really a true professional, I don't say anything. If it's people I've worked with for many years and I know they're real pros, I could say, "Hey, look, man, I'm hearing this, is it something that you can address?" Then it's cool, but if it's people I work with occasionally, or if it's a record I'm just doing, that I don't really know them well, it's risky to go back, because a lot of times they take it personally, like, "How dare you criticize my genius?"

Q: So it could be counterproductive?

Yeah, and a lot of times they can't even find the problem; they don't even know it's there.

Q: If they knew how to trace down the problem, they would have done so, right?

Yeah, or if you talk about bass information or high frequency information, who knows what they're hearing, because I don't know what they're monitoring on. Most monitor situations today are project studios with not much in the way of acoustic design.

Q: So you're working now a lot with surround sound etc.; do you have experience with mastering for 5.1? What would you teach someone if you needed to?

I actually deal with 3-D audio—5.1 is pretty far behind me, but mastering for surround—there's a couple of things. One thing you have to check is, when going to a codec (like in a Dolby format or the DTS format), you have to go in and make sure that the codec isn't doing weird things like throwing your back speakers out of phase with the front, because it does that. You have to go into the codec and make sure you know what it is doing, because a lot of them do weird things. Dolby AC3 throws the back speakers out of phase, I know that, and you have to go turn that off. However, I'd say, mastering in surround, you need less compression. You don't have to try and do as much. If you start crushing, there's usually a lot of reverb in the back channels. You have to be able to listen well; you need a very good environment, which not too many people have. Ultimately, you have to check the codecs—to make the codec and listen back and make sure it's true to what you did, because a lot of times it's not, and a lot of people don't know that.

Q: How did you get onto that?

I think by listening. You make the codec file and then, "Wait. What? That sounds wrong." And people don't want to tell you, and you starting querying Dolby, and they say, "What are you talking about? We're God." They don't tell you, but you can open it up; there's like a preference selection, and you can open it up and go through, there's a whole lot of stuff that the codec is doing, and you can say yes or no. We stumbled upon it by listening: "Hey, the back speakers are out of phase, what the hell is going on?"

Q: When you get a mix in that's really bad, what do you do then?

Well, you assess where it came from: if you know you can't talk to the client, and a lot of times you can't, because it's some young guys doing their record, and you know they don't know what they're doing—really, you're just trying to improve it. Like Greg Calbi used to say, "One letter grade—you can only get a mix one letter grade better." Because if you try to remix it by mastering, you're going to make it worse. What you have to do is assess the possible, and a lot of times you have to think differently than your normal mastering processes, and maybe you have to do things a little different than you used to. But you just try to get it one letter grade [better]. A lot of times there's various methods of doing that. I don't think there's one answer.

Q: On the opposite end of the spectrum, have you ever mastered anything and done almost nothing at all, and have you been challenged on this? What do you say to someone?

A lot of times, the client is not here with me, so they don't know what I didn't do, but a lot of times I've done very little, yet it seemed like a lot. Sometimes you can do really tiny little things and it really improves it. So I've never been challenged on it. I think in the current climate you can do very little, but it's because it's so damaged, or has so many issues, that the more you do the worst it gets. So sometimes doing less is actually the best thing that you can do!

I know Bob Ludwig got challenged on some record, he didn't do anything, nothing. He charged them something like, $10,000. And it was probably the right decision. He saved them from messing up their record.

Q: What are the differences between the way you work in attended and unattended sessions? How do you manage client expectations when they're there, and whether they want to be involved, etc.?

Yeah, the worst thing you can have is a client that wants to be involved. Because they don't know what they're doing. Another problem is, they don't know your monitors. They're coming from some less-than-accurate monitor environment that they are used to, and when they come into a mastering session, they don't know what they're hearing. Also, a lot of times, they demand that you do a lot of things for them, and you do them, and they go home and they say, "This is horrible." But I say, "I told you not to do that." And they say, "Well, I'm not paying you because I didn't know what I was hearing."

The worst thing you can get is a client who wants to be "hands on." It's one of the biggest bullshit lines in the world—"We're really hands on." I think there's a current crop of mastering engineers who have tons of gear, and you sit around and you jerk off in the studio. And the client's "involved," and they're like, "I'm really hands on, man, like it's my art." You're like, "Yeah, the mix came in sounding like shit, but let's fuck with it some more." And they go home and maybe they just feel good about the experience, but I can't do that. I'm not audio day care, man; I just can't do that shit.

So I prefer to say, "Look, I'm going to do a version, I'll send it to you, you listen to it, and let me know, and you can give me revisions." You know what, I never get revisions. The only revisions I get are usually, "Can you change the segue between tracks 4 and 5?" Once in a blue moon, someone says change something. Many artists think that they're their own god, and by messing with the tracking guy, the mixing guy, and the mastering guy, that somehow the record is going to be better. Everybody has a studio in their computer, everyone now knows everything, and they got what they wanted, right? They got everyone out of the way, they're in complete control, and they've never been making less money and never worked less on their music! They've gotten everything they asked for, but I'm not bitter!

Q: Mastering seems to be undergoing a bit of a transformation, certainly during the past decade in particular, where you have a lot of people feeling like they have some kind of expertise, and I'm wondering, do you have strategies for dealing with that? As part of your work, is there something you do to manage the client's involvement in the process?

The best thing I can do is I try to keep the client as far away from me as I can. Many engineers say (at least in public), "I want the client there, I want their input." But the worst thing you can have is someone looking at you while you're working, especially someone emotionally attached to the music. Mastering is an objective art form. You can't love your subject.

If you have to have a client in the room, what you need to do is make them think you're doing a ton of shit. You have to put on a show. A lot of times it's misdirection. Because I'm doing tiny little moves, they can't even hear what you're doing; my moves are usually a dB or less. They're these tiny little things. I also work at low volume, so if they're here, number one, it's not loud enough for them. Because I'm listening quieter than they've ever listened, and I'm jumping volume around and I'm doing these tiny moves, so most of the time, they just get really bored. I work that way, and basically they don't know what's going on. But when I turn it up and A/B it, it's, "WOWWW, it's so good!" So that's the way I keep them uninvolved, they can't hear what I'm doing until I'm ready for them to hear it. They hear the difference, and that's great.

Q: In general, since you've gotten into it, how has mastering changed, if at all?

Well, the budgets have dropped, and so therefore, the people with the great experience and facilities are now competing with people in their basement, and a lot of people are just shopping for the lowest dollar amount. I just had an example last week where somebody emailed me and said, "Hey, I'm applying for some grants, I'm doing a KickStarter, we need promotion, I'm on a little label, I'm on a small budget, I need a five-song EP done."

Usually, for that I would charge about $1,000, depending on what they needed. He said, "My budget is $300." He actually had a good strategy; I felt like, business-wise, this seems worthwhile. I was aware of the label, and I said, "Look man, I'll do it for $300, but there can be no revisions. You send it to me, and I'll do the five tunes." Because I didn't have to do any production, I didn't have to make a CD or anything, I just had to do like five .wav files, which means there's no production involved.

I said, "I'm aware of who you are and who you're working with, but no revisions." And he came back and said, "Well, I need to have revisions." So I said to him, "You're actually asking for people to work really cheap, and yeah, you might find someone that will give you revisions, but it's going to suck when it comes back the first time, and all the revisions are going to suck. You're better off having a good job done once." But no, the artist wants control, and they don't want to spend any money. So I said, "Ciao bella, we're done. I offered to do you a favour, and you're telling me you have to be over my shoulder—yeah, right."

I knew the mixes were going to be challenging to start with, but I don't really chase that bottom dollar anymore because it's exhausting. The most dangerous client is the ignorant client who thinks they know what they're doing. Another thing that's going on, and I don't know why this happens, but you'll give someone a good deal, then they'll want it for $50 less because they think they're doing "business." They think that pissing off the mastering engineer is really the way to go. "I gave you a deal, and now you want it $50 cheaper?" Those are the inexperienced clients, and they're the worst ones to work with.

Q: Where do you see mastering heading, given the things you're talking about?

I think the quality of mastering is going to be gone. I was in New York, and I was at a party at Sterling Sound, which if you've never been is the whole top floor of the Chelsea Market. The rent must be a couple hundred grand a month. Now, I know their lease is up in a few years, so where are they moving to? Are all the old guys going to retire? Where is the money going to come from to rebuild a facility like that? MasterDisc imploded, it's gone. The only one left in New York is Sterling, and when that lease is up, what's going to happen? There's more of a democratization, but there is also less of a really elite skill group, because you don't have a place to learn, and then you don't have the money to support that level of facility. It's a bit of a grim view, but the money is just not there anymore.

Q: If you could go back in time and get yourself right as you were to embark on your career, what would you tell yourself?

I think I would have become an artist and not a technical person. I used to write a lot of records in New York, and if I would give myself the ego, which I don't really have, because you have to have one to be an artist, I think I would have preferred to be an artist as opposed to an engineer.

I just really enjoy writing; I mean, I do still write a lot, but I think I would have tried to make it more as an artist than as a technical person, because I already did the technical part, so I'd go back and try something different. Yeah, I wouldn't do the same path again. I mean I loved this path, I had a great time, but I wouldn't want to do it again; I would want to do something else.

Alex Krotz

Alex Krotz is an audio engineer and mixer based in Toronto. He works full time at Noble Street Studios, where he participates in a wide range of projects, working with industry-leading producers and artists from all genres of music. He has worked on projects ranging from Three Days Grace, Barenaked Ladies, and The Roadhammers to Drake, A-Boogie, and everything in between. With his lifelong enthusiasm for music and great attention to detail, he is recognized as professional and knowledgeable beyond his years.

Q: Why don't you tell us a little bit about yourself?

My name is Alex Krotz, and I work at Noble Street Studios as a full-time assistant engineer and staff engineer. I do a lot of engineering for the studio as well as a lot of engineering, producing, and mixing for personal projects, as well. I work on lots of different types of music, different genres, and lots of different artists. I'm in a great position to be able to see a lot of different approaches from different engineers and producers. It's great, because I can take what I want and put it into my style of engineering and production.

Q: You also have a gold record.

I do; I have a gold record on my wall. It was my first one for Billy Talent "Dead Silence"; now, there are lots of other gold and platinum records hanging at the studio that I've been a part of.

Q: Mastering is a part of the production process, but for a number of different reasons, it's been cordoned off. It's almost like a peer review at the end of everything. There are two different approaches that tend to predominate in mastering. One is the transfer tradition, which is the very early tradition. A mix engineer gives something to the mastering engineer and says, "Make sure it transfers optimally. Don't mess with my balances too much." Then there's the cutter tradition, where a mix engineer works with a mastering engineer, sends a mix, and is okay with tinkering. In general as a mix engineer, let's say you're mixing and you're sending something off for mastering: what do you think is going to happen when you send that track off?

When I send it off, for starters I know that it's going to get louder. It's going to be clearer, and it makes the track have a bit more energy. It sits a lot nicer in the whole balance of everything. That's what I like when I get a master back. I like it when it's basically the same thing I mixed, because I like my balances, but it just tweaks it a little bit to put it in the right place. I don't like when I get a master back and it's completely different. If I wanted it that way, I would have mixed it that way. I do like a little bit here and there, so if it needs it, great, but if it doesn't need it, I'm very okay having it come back and literally being louder. If I'm sending something to get mastered as a part of a full record project, then I expect the mastering engineer will make each song sound like a part of the same record, in terms of just overall low end/high end balance, that type of thing. Again, very slight variation to my mix, because I still like what I did, and I did it for a reason.

Q: Are there any things in particular that mastering engineers do that you routinely ask them not to?

Not really; I'm pretty open to them doing their thing to achieve what they hear. If they think it needs something, or a specific technique, then great. I don't like when they really change it that much, unless it really helps the track or the whole record in a great way. They know lots of different genres of music, and so it's nice sometimes to have that extra ear on it at the end. I know there are some techniques, like mid/side, and I'm open to hearing it if they think it's worth doing, since it's another part of the creative process. Sometimes it works; sometimes I don't like what it ended up doing to my mix, so it ends up not working, but I'm open to the exploration.

Q: You're not necessarily expecting it, but you're open to a mastering engineer doing some work on the mix.

If they think it needs it, then great. Subtle things sure, or to make it fit better into the whole scope of a record, then I'm open to it.

Q: How did you first become aware of a mastering process?

By working with different engineers and them having to get stuff mastered. Asking them, "Well, what is this mastering thing you have to do?" And them saying, "We've got to make it fit with the rest of the record, make it louder and clearer." That's how I got into it when I first started out. I've never called myself a mastering guy. I know a lot of mixers who say, "I do mixing and mastering." I don't say that. I'm not mastering. If you want me to master something that I mixed, I basically put a limiter on it. I make it so when you put it on your iPod, you don't have to turn it up all the way to hear it, but I don't label myself as mastering at all. That's something else, and I like to keep it that way and am fortunate enough to be able to maintain that.

Q: Do you find that that's common?

It's interesting. The more professional clientele that I work with are a lot more about getting the mastering guy to do their thing. Let's do what we do and make it the best we can, and then give it to a mastering guy. There's not a lot of, "Let's fix it in mastering." It's more like, "Let's make it perfect so the mastering guy doesn't have to do a whole lot and has something great to work with." That's common on the professional side of things. For the more amateur people working in their bedrooms and people working in smaller studios, there's a budget difference, and it seems to be harder to justify paying a mastering guy. There's a lot more of those engineers and those types of projects that have the mastering built into the mix. Or the same guy does it at the end and listens to the whole record and masters it to make it fit. Those are the type of people that do the mastering as well, whereas the big guys, not as much.

Q: It seems that the more professional you are, the more experience you have working with mastering engineers and [being] aware of things they can do for you.

That's probably the difference, because the lower guys have never really had that luxury of a separate mastering engineer or really know[ing] what a good mastering engineer can do for the project. Those are also the guys that are doing all their producing, their mixing, their engineering, song writing, and playing the instruments.

Q: You have your basic mix, you're going to submit your mix for mastering, and you have someone in mind. What are the ways you finish off your mix? When you submit the mix for client approval to begin with, you slap a limiter on, right?

Yeah.

Q: Why are we doing that?

One of the reasons is literally so they don't hear it and go, "Why is it so quiet? When I put it on next to my other record, why doesn't it sound the same?"

Q: They interpret that as a deficiency in the mix.

Exactly. They think the mix isn't very good because it's not very loud. The other reason is you want them, and yourself, to hear a bit of what limiting and that stage is going to do to your mix. What is going to end up happening to a point? Obviously, it's not going to be the same, but when you limit something and compress it like that, it does something to the mix. You want to see what it's going to do, because sometimes you hear things that you would change a bit. Like sometimes when the whole mix gets compressed, the vocals become obviously too loud. Not fine-tuning the limiting or anything because that's what the mastering engineer is going to do, but it does give you an overall sense.

Q: I guess you're just inflating the dynamics of the mix. We get a lot of responses where the client doesn't really understand much about mastering. They listen to a quiet mix and think, "What's wrong with it?" That's a really good point—hearing what the mix is going to do when it is inflated.

Part of it is for them, and part of it is for me. I'll put it on and go, "Oh shit. Let me spend another five minutes on that. There's a problem."

Q: Say you have client approval, and you're now going to send the mix off for mastering. What are the steps you take? Do you try to always get it pegging at −6, −8 or some area?

I don't know the number of where it is exactly. I know where it looks like on the meter. Basically, I'm trying to get it as loud as I can without clipping, but I'm not too concerned with the exact numbers. I've built my whole mix around a certain level. When I mix, I know what I'm looking for. I mix at a certain volume, so I know what the kick drum is going to sound like. I know where it should be that it's not going to overload when I put everything else in. That's just from doing a bunch of mixes. That's what I hear and learned to do. Just gain structuring correctly and making sure it doesn't peak on the master. Besides that, I'm not going, "Oh man, I have to put it at −8 or −6." That's my mentality, and as long as we're not completely riding zero the whole time.

Q: There's not any special routine that you'll go through?

Not really. It just happens in the mix for me.

Q: What you're talking about when you are aware of headroom is that you don't have to be mathematically aware; there's no magic number. You could see mixing and mastering as part of the same broader process, just in a different stage. Once upon a time, the mastering engineer was in the building, with the mix engineer in a different room. When you get a master back, what are some of the concrete things that you've marked back to mastering engineer and asked them to change?

They're trying to make it the loudest they possibly can most of the time, and I get that. My notes are more like, "You changed the balance on this or that a bit too much for me," like if I had a massive kick drum and that was the whole point of the song, and they toned that down too much for me to gain extra headroom, then I'll make a note how that big kick was my whole intent. I'll sometimes send things back as well like: "Overall the mids are too scooped and playing with the guitars too much," and just overall things I hear that they did that I want toned down. Sometimes it's the techniques they used that cause the problems I'm hearing, but I don't know what they did exactly, if they used mid/side or something else, so I just tell them what I'm hearing that I don't like; then they should be able to know what they did to cause it and know how to fix that.

Q: I find this to happen a lot in EDM. What I'll often get is rather than notes on the master, I'll often get the clients saying, "I'm going to send you a different mix, I'm going to lower this and switch that out and change my high-hat tone and run it through the exact same line." Do you ever get that?

Not as much in the rock stuff I do. However, the EDM stuff or the more electronic genres makes more sense for those type of notes. You hear the master back, and you like it, except for the high-hat is too big, then you literally click a button and find a different high-hat: fixed. However, when it's rock music or more live instrument-based genres, it's hard to change the high-hat because I recorded drums, and I have 40 mics of stuff that are all contributing to the high-hat sound (for example). Your mix is your mix, and that's what your high-hat sounds like. Its less about one or two elements that can be zoned in on, its more about the overall sounds in those types of productions. So in those cases, it's not my mix that's the problem, because it sounded good before, so it's something the mastering engineer just needs to adjust on their side.

Q: Have you ever engaged in stem mastering?

Not really. My mix is my mix, and that's how I like it. I do print stems, and I might have sent one once for somebody, but not a common thing for me or people I work with. Whenever I print a mix I always print an instrumental, an a cappella, and vocal-up. It is common for me to send the vocal-up mix; I'll say, "I like what you did, but the vocal is too quiet, and that's my end. Here's the vocal-up, and do the same thing." Or I'll just send both and say, "Here, do both, and I'll pick afterwards if I like the vocal-up better." That's the closest thing I've ever done to stems for mastering.

Q: Let's talk a little bit about compression on your mixes. The most common complaints are that the mastering engineer just compresses a lot out of my mix. Have you ever experienced that?

Not really. I have heard a couple of things where I'm listening to it and it ducks a little bit when my kick drum hits. Maybe I'll write that in the note, but besides that it's generally pretty good. I've had a good run of stuff, I guess.

Q: There's this pervasive belief that the mastering engineer is going to compress this to hell. We're having a difficult time finding that in reality.

I think that comes from the producers, because they don't know what's going on. They think that is what happens because that's what they hear. If you read a lot of online forums, there's people who are going, "This sounds good." What are you talking about? It's like where you read on forums that the Fairchild compressor sounds so good on bass. Have you ever actually been in the same room as one? I feel like it's a little bit of the same thing with mastering, where they take to the internet and start saying, "The mastering engineers compress the shit out of everything." Have you ever actually sent a mix to a mastering engineer? I think that may be where the rumour comes from, because generally I don't hear the compression much. Sometimes they barely even use compression, and it's just limiting or clipping. Whatever it is, it's transparent and it sounds fine. Maybe it comes from amateur mastering guys who don't know what they are doing, really.

Q: As a mix engineer, what do you wish mastering engineers knew about the mix process?

I've been lucky to work with mastering engineers who know enough, and know that what I mixed is what I want it to sound like, so don't mess with it too much. Luckily, the ones I have worked with already know, but that would be the one thing, I guess.

Q: One of the things that comes up is they're constantly scooping out the bass.

Because bass take up a lot of headroom and energy, I can see how that would be something that happens. I worked with good guys. I'm in a fortunate position to be able to say that I work with good engineers and mastering guys. A lot of mixes I get back from people and projects I've been working on are almost always bang on, anyway. I know of projects I worked on way back, now that I hear them back now it's like, "For fuck's sakes." I like it when mastering engineers respect your mix. Sometimes they don't; in the few cases, they change it too much. Like I had a big kick drum: leave my big kick drum. If it compromises the master, ask me about it and go, "Hey, I can only get it so loud with this kick drum. Would you like it less kick drum, but louder?" I would say, "No, I want it this way," or "Okay, I didn't think about that, let's try that." See what that sounds like. I don't like it when they alter it too much and start changing the balances of stuff. Just fix the issues of what happens and make it sound nice. That's what I want. I want it pleasing, not totally different, because I already did what I want as a mix engineer. If I have concerns, then sometimes I'll send a note to the mastering guy and say, "Hey, let me know what you think, because I do respect your ears." Sometimes they'll say, "It sounds great" or "If you wanted to do this, then you might not want that big of a kick drum." I'll go back and listen and think about it. I do respect their ears and their thoughts, but I also want them to respect mine and what I've done.

Q: It is almost like a quality control and like a curation thing, where the mastering engineer is supposed to have enough experience with particular aspects like spectral and dynamic balances. They know where it's supposed to be. Their job is to adjust and add a little bit of varnish.

Yes, they add a little bit of coating on top to make it nice and elegant. Also for record projects, they make the whole record sound like a record opposed to a bunch of songs mixed by different people. Sometimes you mix a record over a year or a couple months. At the end, you want the mastering engineer to make it sound like one nice beautiful project and put a bow on it. Quality control is a good way to put it.

Q: Do we need to educate mix engineers better with respect to dynamics, or do we leave that part of the process to the mastering engineer? We're encountering that mastering as a service industry may well be closing. There will still be people that master, just like there are still people who create buggy whips for the Amish community. With this workstation revolution, Pro Tools and the fact that a lot of people are working on laptops, that maybe that mastering is receding and amalgamated into mixing. Do you have any thoughts about that? I know you're working at a level where it's a professional operation. I'm sure, at the beginning of the project, the producer knows who they want to work with for mastering. When it's not that situation, many of us noticed that the client says, "We'll just pay someone $1,000 and we expect it to be everything from beginning to end, including mastered, out of cost consideration." As that expectation becomes more and more common, do we need mastering?

Some amateur mix engineers maybe need some lessons in dynamics, but the good mix engineers, we mix the dynamics as we want them in the song, then the mastering just adds to that a bit, giving it a little added bonus type thing. In that regard, mastering engineers just need to know what type of client they are dealing with, I guess.

I'd say, in that example, I'm part of the community that still wants a mastering guy. I agree that it seems to be closing a bit with the workstation revolution, unfortunately. I think part of that is part of the amateur client's point of view; after they hear the mix engineer's reference master, they don't realize or just have no idea what the mastering guy is doing after the fact. I think it comes from the fact that they have no reference about how much better a project can be with a mastering engineer, because they have never put the money forward to do it. It's a loop: they don't know what it does, so they don't invest in it, and because they don't invest in it, they don't know what it does. They also have no idea sometimes that they need a good song to be able to have a good production, to be able to have a good mix—it all leads into each other. One thing that does happen sometimes is mix engineers hide the mastering cost in their own cost, so the clients, who are paying, can't dispute the cost. If you want me to mix your project, then I want it mastered by this guy, so then it's going to cost this much for this whole package. However, for the really budget-conscious clients, they have mix engineers who don't do that, because they don't want to have their cost appear to be more than it has to be. In the end, it just hurts them anyway, I think, because their mix doesn't sound as good as it could if they got it mastered and didn't do it themselves. It is definitely dependent on the level of clients and people involved in the project. High-end clients do want mastering guys; smaller clients don't see the cost benefit because they just don't know and can't hear the benefit, and if they have the opportunity to cut the cost, mastering seems to be one of the first things to get completely cut out. I wish it wasn't that way.

Alastair Sims

Working with a number of Canada's largest bands (Rush, Tragically Hip, Three Days Grace, Barenaked Ladies), Alastair has been able to work alongside and learn from some of the best engineers and producers in the business. Earning Platinum and Gold records for his engineering and editing with these artists, Alastair has recently been working with EDM and pop acts Loud Luxury and Nikki's Wives, gaining recognition with over 10 million spins on Spotify. Alastair has composing credits on the *Charles Bradly: Soul of America* documentary as well as the forthcoming *Freaks and Geeks: The Documentary*.

Q: When did you first become aware of a mastering process?

I became aware of mastering fairly early on. When I really started getting into music, mastering was like this Ivory Tower process that fixes everything. Everybody was talking about mastering and not really talking about

118

mixing or recording. They wanted everything to be mastered super loud and by the best person. I was intrigued by the mystery from then on.

Q: What did you know about it?

It made a song as loud as possible and by extension made it better. Later on, I went and visited a mastering studio in Toronto called João Carvalho Mastering. He A/B'd a premaster and postmaster. I was like, "Oh, now I get it. It can make stuff sound a billion times better." I couldn't even tell what he was doing, but it sounds amazing. At that point I realized that there is an art to it. It's not only making stuff louder. You can actually do something with it.

Q: My first experience was in the late '90s, and mastering was what the guy did to the digital files that we rendered from tape to do a red book thing. There's these two traditions in mastering. It depends on where you are in the world. One of them is the transfer tradition, where the mix engineer gives a mix and says basically this is what I want. Make it louder and inflate it but don't mess around with it too much. Then there's another tradition that's kind of like the cutter tradition, where they give the mix and know that the mastering engineer is going to have some input on the balance. It's not necessarily huge, but there will be some moderate changes made. In your opinion, which of those two do you tend to work with?

Transfer, for sure. I don't want people to mess around with the stuff I'm doing. If you're getting me to produce or mix a song, whatever project it is, I don't want an outside party (the mastering engineer) to change what the artist and I have recorded, developed, and made. I want them to bridge the gap between the raw mix and a finished delivered song. To me, that doesn't mean drastic changed or mastering from stems: it's light changes to make the mix translate in every situation.

Q: Right, respect your work as the mix engineer. People are divided about that, and I wonder if it has to do with experience and being insecure with your mixes. I find that the types of mixes I get are from people who are starting out and prosumer, generally. They're very dependent. "What do you think? Is that kick okay?" They don't know that it's fine. In your experience, have you had a mastering engineer mark back with notes and having input on the mix?

I'm not closed to it. I'm certainly open and would encourage having communications with the mastering engineer. I've never had any reach out to me for changes. That said, if I was contacted about something in a mix, I'd listen and check it out. I'm not perfect, and if I'm working with them, I would trust them enough to know when something might be up.

Q: In a related way, you and I once had a conversation about mid/side processing.

In rock, you have guitars panned hard left or hard right; in pop, you have a lot of vocals that are panned hard left and right, and in both, lead vocals, kick, and snare dead up the centre. I spend hours trying to get that balance right. If when mastering you go and crank up the side, then you're turning down my lead vocals and drums, completely changing the balanced that I've worked on. That would be why I tend to shy away from mid/side processes.

Q: I think a lot of people forget that if you are messing around with the sides, particularly in LCR mixes, then you're actually messing with the levels of certain tracks. So when you're preparing your mix for mastering, are there particular things that you look for? Do you mark to a particular amount of headroom?

I'll take a look at it; if I've been the one mixing, I'll have been watching the whole time, so I know it won't be clipping or too low or anything weird. While mixing, I have my mix bus processing on all the time, and I'm printing to two-track simultaneously, one that is the output of the mix bus and one that is going through some sort of mastering chain; 99% of the time while mixing, I'll be listening to the "mastered" version. It's nice having both printed so I can send files to the artist or the mastering engineer at any time without having to reprint anything. Usually I aim to peak around −3 dBfs; if I'm achieving that, then I don't do anything else—I send off. I know what the song sounds like when it's limited, so I'm not going to be blown away when it's sent back and some transient are gone. I leave all my bus EQ and compression on because that's something I've been mixing with and into. When mastering engineers say, "Don't ever send a file that has any compression on it." Fuck 'em. That's bullshit.

Q: You go to any mastering website, and they will say, "Before you send it to me, take the compressor off your stereo bus and any other processing." You say, "No that's not necessary, as long as I have some headroom, because I was mixing into it."

Definitely; if you're mixing with anything on your bus, that's your sound. In fact, I would advise against taking processing off your mix bus before printing for mastering. If you've been mixing with a compressor on your bus the whole time, and in the chorus you're hitting it hard so it compresses 3–4 dB, you're shaping the entire dynamic of your song around that. You assuming the mastering engineer can get that exact same effect would be a mistake. I'm not saying they can't achieve that, but that's not their job. They're there to prepare the song to be released. The mix you send should be what everyone has been listening to and agreed on.

Q: I've always thought that if you're compressing for tone and flavour, then it shouldn't have a major dynamic range issue anyway. For the reference mixes you send around, why do you throw the limiter on it?

Everyone is going to be listening in iTunes or on their phone; they'll have either their set volume or something ready to compare it to. Your mix is going to be far out quieter if you send a mix that's peaking at −3 or came straight off a console. If you're running in VU, then it's not even going to be close to coming up to digital 0. You clip it, limit it, or do something, and you get it nice and loud. The number one response you'll get back if you send a quiet mix is that it doesn't have the power of something else. A lot of people use that to their advantage when they send back a mix, and they'll crush it even beyond what a mastering engineer will do. Then you'll like it even more, or you'll get the mastering back and you're like, why does the mix sound better than the master?

Q: Mastering, like so many other service industries and the broader record industry, is closing. A lot of professional places are having trouble making ends meet. One of the things that they're noting is that a lot of producers are saying, "We'll mix and master, too." They slap the limiter on the stereo bus. They say that's not mastering, but it is. We're wondering whether mastering is, aside from the upper echelon of the industry, amalgamating with mixing in the prosumer market, especially with the digital formats, and the transferring from one media to another is pretty transparent. You're not going from tape to vinyl.

It is getting amalgamated; I find it's even being spoken about less now. Possibly because streaming and other digital distributors are using loudness matching to combat volume discrepancies between songs and hyper-compression. We spoke before about how production, mixing, and editing are being smashed together because they are all tools used to create a song. Mastering can be lumped into the same boat: it is a tool used to make a raw mixed file ready for any platform. The important part to remember is the purpose of all of these tools. Why are you mastering? To make something super loud? No, you're are making it ready for any system it'll be played and enjoyed on. You might think, "Isn't that similar to what mixing is doing?" That's why I don't find the amalgamation of all these areas to be a bad thing; they're all processes used to make something. They work in unison to make a song, being done by the same person with the vision isn't a terrible thing, in my opinion. Maybe not better sound quality-wise, but that has never been something that stops great songs.

Q: If that amalgamation is happening, then something's lost in the process?

Of course, you might be losing sound quality, or simply having less ears hear something before it gets released. On the flip side, though, you might be gaining speed, or cutting the budget; you need to find the one that's right for the project. I've sent out a number of mixes that end up being used as the master. The important part is taking the time to step back, think about what you're doing and why. Don't just slap a limiter on and crank it. Ask yourself: what is this song going to be used for, where is it going to be listened to, and how can I compress, EQ, and limit to best achieve that? This should cross your mind at all times; if you're only sending a rough mix to an artist to hear, ask yourself what you should do to make that mix translate. Worst-case scenario, you made your mix better.

Q: Is there an experience you had with mastering that sticks out in your mind as being particularly illustrative of the collaborative nature of mixing and mastering?

The only time that comes to mind was recently when I was asked to mix an EDM track—the artist wanted it incredibly loud. I told the mastering engineer: do whatever you have to do to make it as loud as the reference. We worked a few rounds together tweaking the mix and the master to get the track as loud as requested. Generally, though it's, "Here are files." An hour later I get an email back with the masters, and it's done. If I ask for any changes, it would generally be about the overall loudness of the track or take out any weird stuff they added: EQ, de-essing or some aggressive compression.

Q: Have you ever worked with stem mastering?

No.

Q: That is so common. You hear all about stem mastering. You go to mastering engineers' sites. They always have pages about here's what we'll charge and do for stem mastering. In my entire experience, I've had two people ever even ask about it. Everyone I've spoken to so far in this sort of interview has said, "No, never done it." You're all operating at the upper echelon.

Why would you? You have multiple people mixing it at that point; doesn't make sense. If I wanted someone to mix it, then I would send it to a mixer, then send it to a mastering guy after.

Q: It is weird. It's having a mastering engineer do mixing. Let the mastering engineer be good at mastering and the mix engineer be good at mixing. I'm fascinated, because for all the talk about it, no one does it.

As soon as we got to digital it was easy to create and send the different stems around. With remixes as popular as they are, the stems were getting sent around. Mastering guys were like, "Oh sure, we can do that." It's a product of the music that's being created and the technology of the digital age.

Part Two

Europe

Ray Staff

Ray Staff is one of the most experienced international mastering engineers. With hit after hit spanning all genres of music, his technical skill and creative ability allow him to achieve the optimum result for each release format. In a shootout, Ray will generally be successful against the best from other mastering facilities. Joining the legendary Trident Studios (Soho, London) in 1970, Ray became part of the fledgling mastering department. From the outset, he was contributing to iconic projects: David Bowie, *Aladdin Sane* and *Ziggy Stardust*, and Elton John, *Madman Across the Water*. Ray's skill saw him progress to become Trident's chief mastering engineer. Other classics mastered by Ray include Led Zeppelin, *Physical Graffiti* and *Presence*, Supertramp, *Crime Of The Century* and The Rolling Stones, *It's Only Rock 'N Roll*. More recent projects include the triple platinum No. 1 debut album and single by Corinne Bailey Rae, Fron Male Voice Choir, *Voices Of The Valley*, The Wolfmen, *Modernity Killed Every Night*, Joe Brown, *More Of The Truth*, The Puppini Sisters, *The Rise & Fall Of Ruby Woo*, Paul Heaton, *Crosseyed Rambler* and Daniel Ward Murphy, *Until The Morning Light*.

Ray is regularly featured in hi-fi publications for his work on audiophile vinyl releases, where his skills receive universal praise. He is also much sought after by labels from the demanding worlds of jazz, world music, classical, crossover, and stage shows. Ray is also building an impressive reputation for surround sound mastering with artists such as Gary Moore, Deep Purple, Band De Luc, UB40 and Alice Cooper.

Q: Did you always want to be a mastering engineer?

No.

Q: So how did you get into it, then?

Sort of by accident. I went to school in East London, and at that time the school I went to had just been rebuilt. There was a very forward-looking music teacher there who decided, because I think he had heard about the American and Canadian schools having their own little radio stations, and this was the era of the UK pirate radios, and he thought, "Oh, we should have a little radio station at our school." We started one up, and in fact, that radio station was the second one in the UK. The other one started a week before us, down in Wimbledon, where Chris Kimsey comes from, and he started at that one. That's how we both basically got into the business, through radio stations at school, and that's actually how we both met originally. I was actually interested in doing TV sound and that sort of stuff when I left school, but I ended up falling into a job at Trident Studios. At the time, you just started at the bottom and you worked your way up. After about a year I was asked to do some work in the copy room and cutting room, and that was it, the detour happened. That's the route that I followed.

Q: When did you first become aware there was a mastering process in record production?

I wouldn't call it a mastering process back then. There were two roles, really. There was disc cutting, which in the UK was very much a transfer thing. There was also the copying element, which was doing tape transfers, distribution copies, editing, pre-editing for cutting and compiling, and all those sorts of things. At that time at Trident, that was not really a vibrant part of the business. The studios were doing amazingly well, but the mastering was relatively quiet, and it didn't have that mastering outlook. It was very much still a transfer process: you put a tape on and cut a record flat.

Q: When do you think in the UK all that kind of changed and started to become a . . .

Well, jokingly, we could say when I came along!

Q: I think we should get that one in the book.

It was about that time when myself and a few other people (George Peckham, Denis Blackam, and Ian Cooper) came along. The United States had already been doing things differently. If you listened to some of the American cuts compared to what was happening here, they were very different. Not always better, but different. London started to get independent mastering rooms, so you had Trident, you had Apple, and Abbey Road was obviously doing third-party business. The main competitor, I suppose, then was CBS, who had two cutting rooms, and also Decca, and give or take, that was probably about it. Then it started to get a lot more competitive, and you had to deliver more. We were looking at what was happening across the pond and elsewhere, and going, "Well, okay, we need to do things differently." That was the whole thing, really; the evolution of mastering in the UK was driven by individuals who just wanted to do a better job.

Q: What was the first sort of things you bought or changed about your processes to achieve that?

The main thing was to try and find equalizers, limiters, or compressors, and all those toys that you could be artistic with, which we didn't have. When I first started, it was a basic Neumann console with essentially bass, middle, and treble; that was it. We managed to get some Pultecs, and we had a Fairchild, which was nice. That was the time when we were developing the A-range modules for Trident. We brought some A-range modules into the cutting room and built our own customized desks. I can't exactly remember all the other bits that were in it, but that was the first stage; we made a customized desk and changed the monitoring. At that time, there was new equipment coming out all the time.

Q: An exciting time to be around?

Yeah, it was; it was very interesting.

Q: Was Malcolm Toft doing a lot of the work then at Trident?

Malcolm was the one who interviewed me and gave me the job at Trident. He was studio manager at the time. One of the technical guys at Trident who was really into building things called Tony Simpson (aka Hercules) was a really good electronic and mechanical engineer. He could build things, including the metalwork, and the kit looked like it had come out a factory, it was so good. He loved building things, so he got into it, and that's why the mastering desk was custom built in the studio, and it evolved from that.

Q: I suppose Abbey Road was doing the same thing with their desks?

I think a lot of the mastering consoles that you see at Abbey Road now are very similar to what they had then. I think they were just a variant on the modules that were being used in the studios at the time. They're good consoles, and they had "a sound," and they do still have "a sound." They were definitely interesting, and Decca as well. Decca had a lot of unique bits and pieces they had built themselves. In those days so much was homegrown: people were building their own compressors, their own equalizers, and their own everything, and experimenting. A lot of that is obviously being lost in the business now.

Q: Who taught you how to do cut?

When I moved into cutting there was a guy there called Bob Hill (his cuts had Bobil in the runout groove) who was doing cutting at the time. Bob got me going and started teaching me some basics, but he didn't have a history as a mastering engineer. He was actually a technical engineer who'd found himself in there cutting records, so I learned so much from Bob. Then, after I had been working for a while, maybe about a year, the then-studio technical manager said—because there were a few things that we needed to sort out in the mastering room—he said, "I'm going to get in someone I know to look over the mastering room and sort it out technically for you, and see where we go from there." That's when I was introduced to Sean Davies. Sean spent about a week with us pulling the room apart and rebuilding it. He then turned up on the Friday evening and said he was going to explain a lot more about how to cut records. Sean had come via the pub, and said something like, "I need something to write on." We used to have this great big black wooden box which we used to put in front of the old valve cutting amplifiers because the cleaner had a habit of bashing them and breaking meters with the hoover. So I said, "Well, we've got that box, and we've got some chalk." So there was me, Tony Simpson, and Howard Thompson (who later went on to work at CBS in A&R, and he ended up in publishing and running Rondor Music). There were the three of us there that evening, and Sean, who I have to say was amazing. He chalked up all this information on the blackboard, and gave us a lecture on how a record should be cut, how it works technically, how the lathe worked technically, and how the room should be put together. We were there for about three hours, and even though he was probably drunk, he was so articulate. On that night, I just learned so much.

Q: A magic night?

It really was. It was one of those nights where you went from down there, to up there. It was like, "Wow!" He's my mentor.

Q: Since digital audio workstations have come along, project studios and projects in bedrooms, etc., has that changed the way in which you approach your work when you get those types of project in?

Sometimes it can be a nightmare. You get project studios that do good work, you can't deny that. It's the same as in years gone by, when you had all the major studios like Air, Trident, and Abbey Road. Occasionally someone would come in with something they had done on their little 8-track at home and it would be pretty good, and other times, of course, it would be absolutely atrocious. It depends on the person driving it.

Q: Do you do anything differently?

We just have to equip ourselves with all the different tools that we can think we're going to need, and each project is different. There's never ever one button that I say, "Well, if I'm doing this kind of work, I'm going to need this. Or, I'm doing that, so I'm going to need this." You have to listen to each individual project when it comes in the door, assess it, and maybe experiment to see what works for both me and, most importantly, for the artist or producer and the other people involved. Everyone has to be happy with it, not just me. So you experiment and figure out what works for that project. It could be a good bit of kit, a bad bit of kit, anything that makes it work.

Q: A dumb question, but how much and why do you still value the analog chain and flying in audio into the DAW?

That's a really complicated answer; I was asked a similar question recently. Maybe I should start by saying, "What do I prefer, analog or digital equipment, or plugins?" They all have their benefits and uses. No two digital equalizers or digital compressors are the same. I never trust anything to be perfect. Everything I use will have some attributes or flexibility, but it will also have its own colourations. Although "in the box" equipment has improved, it's not perfect, and there's still things it can't do that we can do in analog. However, yes, digital plugins will get used more and more. As costs become more of an issue, then more things will be done "in the box." Sometimes there is just no way in which you can substitute the result you get from going through analog, or through a particular analog piece of equipment. You can make that assessment by ear, and if you're an engineer, sometimes you can make that same assessment by looking at the engineering and going, "Well, digital can't do that."

Q: We're trying to look at the connections from earlier on the production train until now. How much relationship do you have with mix engineers? If there are issues with mixes, do you often ask for stems to fix things? Do you often go back and have a discussion with mix engineers when they're mixing? Or is it often things arrive, and you got to deal with them?

Both. Quite often, you'll just get stuff that arrives at the door, and people expect you to suddenly get to grips with their music. The same music they've been involved with for the last three months, six months to a year. You're suddenly expected to go, "Alright, I can understand exactly what you want, where you want to go, and what you want to do." There are other occasions, when people can't communicate in the first instance, we try to find out what they're doing, and how are they mixing? We want to know technically a little bit about what they're up to and what they aspire to, and what they listen out for, or maybe some sort of reference that they want their record to sound like. Sometimes we pinpoint a particular genre or style, or another band or artist's sound. Although you can't make them that artist or necessarily give them that sound, but at least you can start to get an idea of what they think they want. When you listen to the mixes, you can start thinking whether or not any of those things are achievable. Sometimes we will get a mix from a client, either sent in or dropped by, and we'll have a very quick listen. It gives them an opportunity nowadays, when people are working in a more domestic environment and the monitoring is possibly less reliable than maybe it has been in the past, to get a better overview of what their mixes sound like in another environment. They get some feedback, and sometimes maybe go back and make changes or assessments based upon what we found.

There are people you work with on a regular basis. We know each other so well, and maybe they're very professional engineers or producers who know what they've got and want. They walk in the door, and we can sit and chat and make an assessment of the music, and work out a plan or experiment on what we can maybe achieve and how the session might evolve.

In every album session, I find the first track always takes the longest. The first couple of hours might be the least productive, because you're trying to get into the session. If everything's come in from one studio or one engineer, you can often get into a place and, within reason, evolve a template or benchmark of how the session is going to go or of how it's going to work. This bunch of kit works on this session, and then it starts to flow. Other occasions, of course, you get a mismatch, and different kinds of sounds are coming from different places. Maybe it's the same engineer, but in different locations, they can still sound totally different. Each of those sounds may then need to be divided up and another approach used, or a different set or combinations of equipment.

Q: What's the one thing you'd say, not your best work, but the thing that you're most proud of?

There are lots of records I've been proud of. The Bowie stuff: *Hunky Dory*, *Ziggy Stardust*, and *Aladdin Sane*. Led Zeppelin—*Physical Graffiti*—that was an amazing cut. A lot of people come in and actually go, "Wow, that was special," and relate to that. I sometimes get the odd email about it. I've had emails from people in Japan saying, "We've got copies of this! We've got these! We found an acetate from 1970." *Physical Graffiti* was something special, and that was a mammoth task, because that was a double album. We had to cut 19 sets of lacquers, totalling 54 sides. We had to do something like 20-odd quarter-inch copy tapes, and this was all in real time. Typical of the music business, they needed it all done in two or three days, so three of us had to work a shift. Once we got it all set up, and I've got everything set with Jimmy Page, we just went around the clock for about two solid days. It was the longest time I ever spent in the mastering room at one time. It was 36 hours of cutting, just going through the album cutting the 54 sides. I had a cue list of what we had to do, and we did that over and over again. That was a killer.

Q: Without revealing the track, album, or artist, can you tell us your worst session ever? About what went wrong, what were the issues?

I don't know the worst session. I've had sessions where clients have been drunk out of their head. I remember a couple of reggae guys used to come in, and they would roll up such big joints. I don't smoke, but after a half-hour with them we might as well have all just gone home, it used to be crazy. I've had some obnoxious people. I've had a client's dog who walked in who had diarrhoea and crapped all over the floor. Oh, God, there are a lot of them, some of which I can't mention.

Q: Has anything gone wrong technically or musically that you've not been able to deal with? Because there must be cuts that you've made where you're just not happy with the outcome, and you really don't want your name on it. Have you ever had those kinds of instances?

Yeah, of course, I won't say regularly. You get stuff coming in which clients are happy with, but at the same time you don't really enjoy it—whether it's musically or sonically, it's not for you. There are a couple of records where I've cut reluctantly because I didn't agree with the content on some of the material or some of the lyrics, things like that.

Q: Absolutely. Where do you think the mastering industry is headed? I know that's a loaded question, and we talk about this lots anyway. Obviously you can answer however you wish to.

I don't know where the mastering industry is headed. It's going to be an interesting time. It's still going to be governed by what happens in terms of engineering. By that I mean, at the moment, to accomplish what we do, we need a certain standard of environment, a certain standard of professionalism, a certain variety of equipment, and everything else. Of course, as engineering and computing evolves, maybe these things can be dialled in and done "automatically," or at least done "in the box" more than it is now. The thing about mastering is the problem of people understanding mastering. By that I mean what's achievable, what's doable, and what's not. I get very surprised, because I meet quite a lot of people that recently come out of universities and really do not understand what mastering is about. They think it's just a preset on a limiter, and you just make it loud and that is it. They don't understand how we listen to it, and it is the art of the individual to say, "What can I do to this that makes it much more pleasing or enjoyable and gives it the energy that this particular material deserves?" I don't think you're ever going to do that in software. You'll still need people there for a long, long time that are able to make those decisions.

Q: The emotional level of it, it's the connectivity with how it moves you. You're talking about automation [machine learning/AI] of some description, whether it's here now (wink wink) or whether it's coming. It can't be done; it has to be an iterative process by which you understand music by.

It does. You've already got things like so-called cloners, where you can make an FFT analysis of a track and apply that curve to another track. It has an interesting result, and it has its uses, but it's not what mastering is about. Mastering is, as you say, the experience of the individual to listen to it and be able to hear beyond what's just being played and say, "What is it that will improve this as a listening experience?" I can't see how you can ever automate those sorts of things, I really don't. In terms of the financial issues and future careers, then that's very different, because the budgets are changing in the music business, and it makes it harder to succeed and survive in a large facility. Maybe it will become more domesticated or more homegrown. Already one of the big changes I see is the amount of work I get from all over the world—it's not just from here in London. Maybe 15, 20, or 30 years ago, you might have had work from here, the United States, and maybe some work from Europe. In those cases, people would have flown in physically, tapes in hand, to see you, because they thought you could do the best job for them. Nowadays, I get work from every continent in the world, and every country under the sun, and people are sending it across the net. It's great, because for me, I'm getting to master really unusual music, something different and something challenging. It's not just another version of a rock band or a pop band, it's totally different music. It gives me a challenge to sit back and ask what can I do with this, with my background. That definitely makes life both musically enjoyable and challenging.

Q: Stems. Like them or hate them?

A bit of each.

Q: How often do you get stems?

Not massively. We get some stems. I do take the opinion that a mix should be a mix. If you go to a stem, it's because you are rescuing something or there is a deficiency in making that final decision at the mix stage. Will the stems sound like the mix? No. If you render out the stems they won't sound like the mix that's come off the original rig, they will sound different. Quite often it means that if someone sends you the stems, they don't actually know what it sounds like, they're referencing it to what they heard as a mix, not what actually they've provided you as stems, which is something slightly different.

The summing amp is going to be different, or the way they've rendered the stems. I've had some recently which I had to just totally reject, because

I had a stereo mix and it was just so different it was untrue. I had no idea what they were doing. It's similar to an instrumental: if you make an instrumental where you have just muted the vocal out of the mix, where the vocal was working and integrating with the whole mix, the backing track is not necessarily exactly the same as the backing track plus vocal. For instance, the backing track without the vocal might hit the bus compressor differently than with the vocal. You have to be careful with that sort of thing.

I've used stems to overcome some real issues and done it very successfully, so I'm happy to do that. When you're concentrating on a job like mastering, you're concentrating on one skillset. You're applying yourself as a mastering engineer—you're thinking not as a mix engineer, you're thinking as a mastering engineer. If you've suddenly got to add that other skillset and that other role of being a semi-mixing engineer as well as a mastering engineer, that can blur your decisions and thought processes. It's similar to when some people engineer and produce the record; it doesn't always work. They can be an engineer or a producer on the record, but quite often they find it hard to actually do both at the same time.

Q: Similarly, engineering and then mastering . . .

It's the same sort of thing. It's a different mindset and a different skillset. It's not that I don't recommend stem mastering—it is a very useful tool to overcome problem mixes, and that's the application I would look for.

Q: The mixes should ideally be the best they should be.

They should be, yeah. I always reflect back to when I started. A lot of the engineers that I worked with in those early days—whether it was Tony Visconti, certainly Gus Dudgeon, Ken Scott, and many others—they would come in, and they would put their tape on and expected it to be right. If you were to do a change in the EQ, it was primarily a minor compensation. If you had to really start working it, they'd want to know why, because they knew what they had was right and what they really wanted to hear. Whereas a lot of people come in the door nowadays, and go, "Well, I've got this mix, what can you do with this?" I say, "Well, isn't that what you want?" They respond, "I don't know?" They're not sure anymore.

Q: It's often said the DAW culture and methods of recording now mean people don't make decisions anymore.

They don't make decisions. I mean, don't get me wrong, there are some people that do for sure. There are also a lot of people who are just unsure of themselves to (A) make a decision or to (B) make a unique sound. A lot of records are very "samey" nowadays. Back in the '60s, and certainly the '70s, I could put a reel of tape on, and you wouldn't have to look at the box. I'd be able to go, "Oh, that's Abbey Road, or that's AIR, or that's the Trident sound." They had their unique characters, or engineers had their certain techniques where, often, you could just pick out the engineer and the producer's style. That's really much harder to do nowadays.

Q: Well, exactly right you are. How would you say remastering work differs? Especially on something like this [David Bowie reissues] where you're involved originally, and you're working on these things now. How do you feel about that?

It can be fun.

Q: Have you gone back to the original master tapes on this?

On lots of projects we go back to the original master tapes. We always try to go back as far as we can. On this particular project, I think we found the original masters probably for nearly everything, or at least if not the original master, certainly an excellent 1–1 analog copy of the original.

A lot of these particular recordings and tapes have actually held up well. We've had problems with baking the tapes. In fact, what is actually ironic is that a lot of the tapes, for instance EMI tape 815 and 816, have really held their own after 35–40 years. Certain other makes of tapes have not. The tapes have held up well, and we've had to do some baking and a little bit of restoration on them, but nothing massive, really.

How do we approach it? For one thing, on something like this where I may not have written notes from 40 years ago, but I have sometimes a good recollection of what we did and would probably have known the equipment we would have been using at the time. I also may have copies of the original records, too. So we refer back to whether we got it right or wrong then, and if it can be better from what we did before. In this instance, this is me and the producers and communicating and talking with the record companies about what we want out of this. Is it what we had before? Or is it something drastically new? Very much on this project, it was talking to the producers, and for the most part [they] were saying what you had on those tapes is what we want to hear. They've made the decision that the tapes were more or less what they wanted; they were right. They didn't want a modern CD approach, they didn't want heavily compressed and heavily limited records. The compression that's on the tape is what they wanted in the first place. Some of these tracks are very heavily compressed in analog, but it's been done for a good musical reason, and it works. Why add any more if additional compression doesn't make it any better? This particular project is going to be moderately loud for the most part, not quiet, but moderately loud. Within reason, we've used a similar style of approach that we would have used all those years ago.

Again, we worked through it, we captured the tapes, restored the tapes, sent copies to producers, asked how it sounded, got feedback, and amended it if we needed to, maybe once or twice, but it's been a fairly straightforward process of actually getting it back to what we wanted it to be. Now we have to think about what it will be used for, because when we first did it, it was only for vinyl. During a lot of these projects cassettes weren't even around, so it was only vinyl then. Now we're obviously looking at it to go across every single format under the sun and have it archived. If the tapes were to age or to deteriorate in the future, we wouldn't have something decent to fall back on. So we also make a good quality archive and keep good documentation with the archive.

Documentation doesn't always happen nowadays on many digital projects; it's absolutely atrocious. Hopefully with the documentation on the files in storage, someone can pick it up in the future and know what's there and what to use [it] for. Do you need to do something totally different from the vinyl version to the CD version? Minor changes, and there have been a couple of those things, but nothing horrendous.

Q: Are you doing slight changes to EQs and things between the mp3 versions, the vinyl, obviously? Are you doing considerable tonal changes, or just getting it right from the format?

First thing in this project, we've remastered everything at 192 kHz—well, nearly everything, there have been some exceptions because it just happens that they were made at 48 k or 96 k originally. If it stayed in the digital domain, it may have stayed at the same sample rate. Some have actually gone through analog and come back up to 192 k, because they've got to marry with other files. We've tried to set 192 k as our quality and technical standard throughout, as it's a really good quality. For the most part, we've been able to use the 192 k files, although some of them needed restoration for clicks and dodgy edits. Because of the ageing process on tape, we can't always work directly from the tapes. We would all love to be able to put on a reel of tape and to be able to cut straight from that reel of tape, straight to the vinyl. That's getting much harder to do now because tapes aren't always in a good enough condition to do that. Some of this project has had to be cut from the digital files because the analog was not in a good enough condition. All the other formats are extracted from the 192 kHz files. For mp3s or digital online delivery, that sort of thing, and because it's not overly loud, and not smashing the end stops, there really isn't much of an issue going down from 192 k to a lower bit rate.

Q: Just use what you've got and repurpose it.

Pretty well for mp3s and AACs. We have listened to some of it in that format, and it's really hard to compensate properly for something that throws musical detail away, so there's not a lot you can really do. The main thing is to not overdrive it, because that's what makes it sound worse.

Q: The loudness wars . . .

The loudness wars, that's the interesting thing at the moment, with all this EBU R128 standards being introduced for TV and broadcasting. Well, that and recent Spotify [and other online providers'] loudness specs. Will it influence what we do in future?

People often wonder why you're working at higher sample rates. A lot of people say, "Well, does it make any difference?" For me, it makes huge difference. Transients recorded at 192 k are so much better with the air and the space. There are lots of good engineering reasons why higher sample rates sound better. One track that we were working on had a really hard guitar which had been driven very aggressively through an old reverb plate; it had lots of harmonics. When we were trying to make a quick listening copy onto a CD from the 192 k files, the track was sounding very strange. By chance, I had a third octave analyzer up on the screen, and it's a very broad band and goes up to something ridiculous like 100 k. We could actually see musical energy going up to about 60 k off a 1/4 inch tape. You thought, "Okay, I know I can't hear it, but when I remove it, something is different, and that wasn't right." I can't remember what we did, but we had to filter it in some way so that when it went back down to CD quality, it sounded okay. It would never be the same, but we helped compensate for it somehow. That to me was a real eye opener, to actually go, "Okay, a 1/4 inch reel tape can actually record those harmonics!" You could see it so blatantly on the screen.

Q: Now that's fascinating. So it just proves that principle from a 1/4 inch tape as well. Fantastic.

I can't remember if it was Dolby or not. It was definitely ¼ inch.

Q: What is the one piece of gear, the lathe aside, you couldn't do without if you were to set up a room tomorrow? Or haven't you got it?

The most important piece of kit is the monitoring. It's about knowing what you're listening to is right. There's no one necessarily favourite equalizer or compressor that I cannot do without, because I use various pieces of equipment. The most important thing for me is knowing what I'm listening to is being reproduced properly. I can hear what's happening properly, then I can make decisions about all the other equipment I want to use.

Q: Clever answer.

Do I get 10/10 for that one?

Q: Of course! What things outside of the day-to-day perhaps instead of day to day mastering are you involved in currently?

There's so much I've been working on at the moment. I'm really interested in what wordclock and jitter is doing. This is the kind of thing that the home studios can't really, don't always have the time or knowledge to look at, and it's not certainly not researched enough. People don't realize how important clocking and jitter really is, so that's one of the things I've been working on.

The other thing of course for the future is metadata, which is becoming a big issue; hence Barry [Grint] especially was very involved with instigating the ISRC code into the EBU .bwav spec, and there's other ideas in the pipeline that are being worked on. So, there's a lot happening that we're looking at. There have been some interesting things going on lately.

Q: We missed any possibility of interviewing Doug Sax.

That's a great shame. He was one of those people you look back at as an innovative mastering engineer. Back in the '70s, I would say it was obvious if an album was a Sterling Sound cut or a Doug Sax cut. You could sort of guess where the cuts come from, which is a little bit harder today. Certain people do have a style, but Doug had that unique quality to his stuff, and if your material suited his system, then great, you got a great result. He was one of the first ever real independents doing that, so hats off to him.

Q: How do you listen, how do you hear?

I think it's the training of the ear. It's one of these interesting conversations I had the other day. I and a few other people are quite interested in this clocking and jitter issue. One of the things that we were talking about was, "How do you define a specification for jitter?" How low should that jitter should be, and what can the human ear perceive? At least then you would know that that specification has some validity or benchmark. Then, even if someone else says, "Well, I actually prefer this," and the jitter is worse, and they like it because aesthetically it happens to give them what they like, then that's a different conversation. At the moment, we can't actually say, "You know what, you're in the wrong place," and making the decision for the wrong reason. That sort of thing is very interesting for me, because people don't know what jitter sounds like. Part of the same conversation was about how many people have actually heard pure analog recordings? Quite often if you went to a lot of studios now and certainly a lot of colleges, and if you asked them whether or not they had listened to a half-inch master played back, they would say no. So, they don't know or have knowledge about what sound could be, because it's being restricted by the facilities and the equipment that they are using today, so they can't always accumulate that historic knowledge of sound that people of my era have got. Their knowledge is often restricted to digital audio. I think there are some big educational issues, being able to go back and say to people, "This is what sound can do, or how it can sound." These are the issues that you can pick out. In fact, if you go to Dave Hill's website, Crane Song [also interviewed within this book], he's got a bunch of files with various levels of jitter on. It's a real challenge.

This is an educational thing, which I think is very important. It's not only people knowing what we do, the anecdotal things which may be entertaining, but if people like myself are going to pass anything on, it's what does the next generation learn from us that we think is important, and what listening skills should they be developing, not only listening in a musical sense, but in an engineering sense, because there are different ways of listening.

Well, how do you do a listening test? Most people would say doing a blind listening test is the only valid way of doing a test, and I can understand the validity of that argument.

However, I've learned that in a blind A/B test you can actually listen differently. So, in other words, if I sit here and I've got the speakers up at a sensible level, and I'm listening and concentrating, I hear a sound and I will pick up on certain issues. Sometimes when you're not directly listening, quite often you've turned down the volume and you're just working on something else, say typing or whatever, and the music is playing in the background, you can pick up on little changes, and quite often you would pick up on something actually quite serious, and you'd go, "You know what, when I was sitting there concentrating I couldn't hear that issue. But when my mind is not 100% focused on it, it seems to register in a subliminally way?"

Q: I know exactly what you mean, it's interesting, that. Sometimes I'll miss something because I'm focussing too much on something else. I'll miss something really obvious like a click, because I'm trying to hone into something else particularly when I'm listening.

Having come from a vinyl background, sometimes I don't even worry about the odd click, whereas other people would panic about it, and that's just a different background and that's because I've come from that era. As you say, you're concentrating on the EQ or the generalized sound, and you're not thinking about clicks, and a click will pass you by because you're focused on that particular job in hand or that particular issue.

Q: I'm glad it's not just me, though.

The idea where you can walk into a mastering studio and expect that the mastering engineer is going to get it right the first time, is pretty naïve. Would you walk into a mixing room and expect to walk out with a perfect mix on that day? No. Quite often, you'll do recalls. Quite often, you'll have totally the wrong approach, and you'd have to go back and do something totally different; you can do that in mastering. You can completely bark up the wrong tree. We're human, we can have good days and bad days.

Q: That's good, because that dispels the so-called "dark art" of mastering really.

It's a different art, really. We're just humans with similar talents to other people, and putting them to good use in the right way, but we're still human. On some days you'll be firing on all cylinders, and others you won't, and you might not even know it until the recalls come.

It goes back to the approach to the project: do you sit back and master a whole album, or do you actually go and do some experiments on one or two tracks, and sit back and reflect? Then come back and think about doing the album, because if you dive in and you have barked up the wrong tree, then you've wasted hours.

Q: Is the album dying? A lot of the work, lower down the food chain where [the authors] work, is EPs and singles. Singles or EPs, where the whole concept of flow across an album is lost. I mean are you finding that at all, or is it still all very much album work for you?

Not with new material. There is a lot of album work, but there are people who are just doing EPs or singles as a step towards an album. I've even done it on projects where we've done three EP's in a row and then they put it together as an album. I think musicians often like to be able to say, "We've got a complete album," because they like to have that collection of songs, because it means something to them musically. Is a single really representative of what an artist is, or what she's doing? No, it isn't really, but people will cherry-pick singles. It doesn't mean that they will really know and understand the artist and understand their music.

Q: It worries us a little bit in the sense that people just master single tracks and then throw them together on an album thinking they're mastered, rather than having them re-looked at. That's the experience we are finding.

That's the problem I have sometimes. Someone wants to do a single, and they will push you in one direction maybe for level or whatever. Then they give you all the tracks on the album, and you listen to what you did on the single and say, "Well, that's not appropriate for this as an album, it just doesn't fit. You can't apply that character to the rest of the songs. You need to do it totally different to make this a really good listening experience as an album." That's a discussion which is quite hard sometimes. It's the same thing with compilation albums. You get people who just send all their stuff in, and they have all these mastered versions and go, "Can we make that all live on the same playlist?" It's not always easy.

Also, there's a knowledge problem in the people who actually do the compiling and put the albums together. They only think about it sometimes in their head musically and won't think about how it functions together sonically. Even if those two songs work together musically, they may not function sonically together. If you try to actually adjust one sonically to marry it up with the other, it may actually be to the detriment of one of those songs. People have got to think about this when they are doing compilation albums. Do they sonically gel together? Are they actually going to flow when they are compiled?

Denis Blackham

Denis Blackham began mastering in 1969 at IBC Studios in London. He later worked for Polygram, RCA, The Master Room, Nimbus Records, Tape One, and Porky's. In 1996, he set up his own home studio in Surrey, England, and when he moved to the Isle of Skye in 2002, he changed the name of this studio to Skye Mastering. Denis has mastered records by the likes of Antony and the Johnsons, Brian Eno & David Byrne, Wilson Pickett, Otis Redding, Yes, Black Sabbath, the Bee Gees, Elvis Costello, Dolly Parton, Jethro Tull, Cabaret Voltaire, Alien Sex Fiend, Tears for Fears, and too many others to count. Denis semi-retired on May 5, 2014. He still masters, just not every day.

Q: How did you become a mastering engineer?

When I was about 14 or 15, I got a book out of the library called *Disc Recording and Reproduction*, and for some reason I never took it back.

It fascinated me, and I still have the book, reading about how music was recorded and made into a record.

If I go back to 1963, I was about 11, and my Dad asked me if I'd like a record player, so he bought me one. The first single I ever had was Dusty Springfield, "I Only Want To Be With You," and the first album I ever had was The Troggs' first album. Because I was still at school, my dad bought me singles every now and again; it was The Beatles, The Shadows and stuff like that back then. Two or three years later, he asked if I'd like a tape recorder. I said yes, please, so he bought me a basic Grundig reel-to-reel tape recorder. I was always fascinated with records and used to look at them in the light to see the groove. On some discs, you could see where the groove started at the edge of the record, where the cutting stylus dropped in, and on some you couldn't. You've also got the differences of the groove "packing" depending on the sound and everything. A lot of American discs were fixed pitch in those days, so you've got this constant groove pitch to the end, and some had slightly different runouts to others, things like that, all these things fascinated me, and I wanted to learn more.

So as I said earlier, I had this book filled with loads of interesting information about how records were made, and I just felt that was the sort of thing I should have as a career. I didn't know how difficult it was to get a job in a recording studio. I think at the time, I wanted to be a recording engineer, so I looked up "Recording Studios" in the British Yellow Pages telephone directory, which had all of the businesses listed in it. I wrote off to about six studios and got the standard letter back from all of them saying, "No." So I just carried on working as a silk screen printer. About a month later I received a letter from IBC Studios, which was in Portland Place, London, one of the best studios at that time; it sadly it doesn't exist anymore. They said "We haven't got a job in the studio, but would you like to be a trainee disc cutting engineer?" Of course I was interested in how records were made, and I'd been reading all about disc cutting in the book. There were photos of a cutting room, and it was actually one of the cutting rooms in IBC—what a coincidence!

So I went up for this interview with Mike Claydon, who was the studio manager and an excellent recording engineer. I had the interview, and he said, "Thanks, we'll let you know." I thought, "Oh well, that's probably the end of that." I asked if I could possibly have a look at the cutting room? So he took me down there, I walked into this room, and I recognized it from the book, so I said, "Oh, this is that, and that's that." I think that probably got me the job, and I've never looked back.

Q: So once you're in the door there, how did you learn to master?

The main cutting engineer at IBC at the time was a guy called Brian Caroll, and he basically taught me how to cut. I was never a "tea boy," as such, but in my spare time I would demagnetize reels of new tape for the studio. IBC had separate mono and stereo cutting rooms. They were both Lyrec lathes with Lyrec tape machines and Tannoy speakers. In the mono room we had one Pultec EQ and one Fairchild Compressor. In the stereo room, we had just the EQs on the Lyrec stereo tape machine and an IBC-built compressor, so it was all very basic at the time, as were most cutting rooms in Britain for that period. American cutting rooms were a bit more advanced then, because it was a different situation. In Britain, disc cutting, or mastering, as we call it now, was generally felt to be a manufacturing part of the business. The project budget was spent on the recording, and disc cutting was classed as a manufacturing process, so record companies didn't spend a lot of money on the cut, whereas in America, they thought of it more as part of a studio situation, so it received a greater part of the budget. The cutting rooms had more equipment, and the engineers were more inclined to be able to mess around with the sound, because they could spend more time and charge a higher rate. In England, we would usually make the best straight transfer we could from what we were given; we didn't really need to do very much at the time and weren't asked to, either, but in time, equipment improved, new cutting rooms similar to American ones were built, and labels realized a good cut could make all the difference. So after I'd learned the basics of disc cutting, I was given the freedom to do what I wanted, so I could start working directly with producers to get the best sound on disc from the mixes they brought in.

Q: Was there really that Atlantic divide?

It was exactly that. At the time (late '60s to early '70s), I used to do cuts for Atlantic Records in England. Basically, we would get a tape from America made from the original master tape or a copy, so there was no EQ. It was usually 15 IPS quarter-inch, and sometimes it was only 7½ IPS, plus in 1969 there was no Dolby noise reduction. I would usually get the tape, plus the American pressing to play through, and work out what I had to do to get the tape sounding similar to pressing. At that time, no one had thought to put a feed on the output of the cutting desk to record the EQ's compression, etc., which was a wonderful thing when that finally happened.

Q: So even what you were delivered was different?

Oh totally, totally! No wonder a lot of people, when CD came out, would say, "Oh, the American version is much better than the British." Well, it probably was, because generally, we were working from a second-, third-, or fourth-generation copy tape with totally different EQs. We had no idea what they did in America at the time, just trying to do roughly the same. Also, in America they generally used Scully lathes with Westrex heads, whereas in Britain, at the time, we were using Neumann or Lyrec lathes with Ortofon heads, and they're totally different-sounding pieces of equipment. Westrex heads generally had a nice little lift around 5 kHz, which made everything sound better without you doing anything.

Q: So then, our most basic question, or talking point. How do you explain what you do—mastering?

It's a question so many people ask, and when we're with friends and acquaintances, or when we go on holiday, people ask, "What do you do?" I tell them, and they still don't understand.

So in layman's terms, mastering is essentially doing the best transfer of what you are given, but on occasions, more often than not, you may need to help that sound to be better. Also, if you think of an album with, say, 12 tracks on it, they could be recorded in six or seven different studios, with several producers, engineers, and different equipment. When it comes to the mastering, that's often the first time all the tracks are put together in the correct running order, and you're able to hear the differences between all of the tracks, so it's the mastering engineer's job to make them all flow, and sound complete.

I might get one mix which sounds great, and the producer loves it, but the following track might sound slightly thinner, the bass drum might be a bit louder, or the vocal might be a bit softer, so I have to mess around with those a little bit, just to get the tracks comfortably sitting together. You need to be able to listen to the whole album, and it needs to flow smoothly from beginning to end.

I'm still concerned when I buy CDs that some of that still doesn't happen; you put an album on and some tracks are quieter or louder, and I don't know whether that's because the producer's insisted on tracks being that way or not. I think in some ways I'm quite lucky, because I've purposely taken myself away from working in London, where producers can come and sit next to me saying what they feel. Because I choose to do it on my own, I put my mark on it and I do whatever I feel, essentially the way I've always worked, and most of the time people like what I've done to their mixes, even though it might be quite different to what they've actually done in the studio at the time, because it has to fit with the track before and after it. I think that is essentially what mastering is, putting those final finishing touches on somebody's project. When I started it was transferring it to a groove, and now I'm transferring it to a digital medium.

Q: As you're doing this, when you get the material, are there things in particular that you listen for, regardless of genre? When you approach mastering, where do your ears go?

I think for anything with a vocal, the vocal has to be more or less the same level and sound throughout the album, because that's what the majority of people are listening to—they are listening to the lyrics; then it's the backing sitting comfortably with the vocals. I'm messing around with compression and a bit of EQ tweaking, just to get the vocal sitting in the right space, and using compression to get the backing to sit with it properly.

Q: Do you have a preferred or particular chain that you typically work through, and could you take us through it?

I do. I look at a lot of interviews in magazines, and I always love to see photographs of other engineers' mastering setups, and I've probably got one of the most minimal ones you could imagine; I really have. I read interviews, and they say, "I've got half a dozen different equalizers depending on what I'm doing and what I want things to sound like," and I'm totally the opposite. I have one pair of wonderful ATC SCM100A monitors which I've had since 1996, and had them upgraded a couple of years ago; I don't have any small monitors; I also have a pair of Sennheiser headphones if I need them, mainly for when I'm de-clicking something, but generally, I use the one pair of speakers, because for me, I know them well and I don't need to switch between large and small monitors. From 1969 until now, I think I've got enough experience to know what comes out of those speakers, because they're very faithful. They work very well within the acoustics of my mastering studio; they're 100% perfect for me. I think I can understand what a sound will be like whether I'm playing it on those speakers or someone's going to listen on a pair of headphones. I think I try to get a sound that's going to be good on an overall basis.

So to back up, I only have one analog EQ, which is a Prism Masalec; I use it, but not all the time. I have two computers, and both are PCs. I would like to use Macs, but unfortunately the software is PC only, so I use PCs. I have them in a separate machine room, so my studio is nice and quiet. I play files from one computer using Samplitude software; the digital output goes through a DCS D/A down to analog, which goes through a Manley Vari-Mu, my favourite piece of equipment for giving me a good basic sound to work with. Even if I'm not using any compression, just playing it through the Manley sorts out probably about 70% of people's concerns. So it goes through that, then the Prism EQ, then it goes into the TC 6000 Mark II up to 24 bit. I use four digital processes on that: I've got a de-esser, and an EQ, a brickwall limiter, and another EQ, and so I use variations on those four. From there the signal goes into the front end of SADiE on a different computer, and I have a Waves L316 sitting on the input. I then use SADiE like a tape recorder, doing all my processing before I record it, unlike a lot of engineers, who master "in the box." If people want things changed dramatically, it might be harder for me in the way I work, because I would have to remaster it again. A lot of people would say, "It sounds fantastic, but maybe track 7 could use a little bit more bass." All I'd do then would be to EQ the mastered version I've done with a little bit more bass, not remaster it as such, just tweak it. That's the way I work, and I'm still here, and a lot of people like the end result, even though I don't have a lot of equipment. I have what my ears like, and with 45-odd years' experience, I know what I'm doing with what's in front of me.

Back in 1996, when I first set up on my own, I was trying out various outboard EQs, and I tried out Focusrite, the Prism, and Manley EQs. I get an idea in my head of what sound I'm after, and I'm thinking, "Oh, it needs this or that." Something has to feel right to me inside before I'm happy with it. Maybe it needs something around 3 or 4 kHz or whatever, that's the way my brain's thinking, so I'll try that, but it doesn't sound right? I'd do that with the Manley, which a lot of people love, but it didn't do what my brain was telling me I wanted, and what I thought I wanted from that particular frequency. The Focusrite was exactly the same, it didn't do what I wanted, but as soon as I went to the Prism, and I did what my brain was telling me to do, it was there, and I thought, "Fine!" So I bought the Prism, and it's brilliant for me; that's the way I do things.

Q: When using your Prism, are there any sort of routine EQ moves that you typically make? Are there areas that you find yourself addressing?

Funnily enough, I generally start off with half a dB around 98 Hz, and half a dB at 126 Hz, with them both set at 14 on the Q, minimal stuff, but it's just a little bit of low end that a lot of mixes I receive need. As soon as I play something, I can tell straightaway if I don't need it, and I just turn it off. A lot of people prefer to have a little bit more warmth down on the bottom end; I like the higher mids on the Prism, too. Maybe it's something to do with the way the Manley valves are changing that bottom end; it could be that, who knows? What I often do, and the way I generally work, is that I'll put all of the files of an album up in one go, assuming they're all the same sample rate, and I'll look at the waveforms. Maybe track six is the beefiest one of the whole album, so I'll master that one first, and then build the rest of the album around it so everything sits properly; that's the way I generally work. What I often do is record that track onto SADiE absolutely flat, exactly as the mix, so that I can flip from the flat to my mastered version at any time and make sure I'm not going too far away from what the producer has done. You need to make sure that what you're doing is actually better than what they've given you in the first place. Sometimes you can go wrong without realizing it, so it's good to go back to the original mix. Maybe part of the original sound is better than what I'm doing, so I will change the way I work, and maybe I'll do it completely digitally, taking out all the analog. Usually, I'm working in the format that I said before, a bit of analog, a bit of digital, and a couple of plugins, that's all.

Q: So, not that minimal?

Yeah, but when you look at the excellent Gateway mastering in America, that wonderful desk, all that gear, and all those bits and pieces they can use, mine really is minimal. If you look at John Dent's place over here, it's another approach. John's always had his own way of working, which is different to the norm. He's got 30 or 40 pieces of equipment sitting to the side of him, just sitting there on a desk. Whatever pieces of gear he wants, he goes around the back and hand plugs everything together into the signal chain he wants.

I went over to Sterling Sound around 1980 to master something for a client and could not believe the set up over there. The engineer was sleeping in the room! He had about four different sets of speakers, and they were all over the floor. What amazed me was, the whole of the right-hand side of the room had windows looking out onto Broadway, and I was thinking, "How can this be acoustically accurate?" Yet they were a fantastic place, and a lot of major albums were mastered there.

Q: You mentioned earlier that you use Sennheiser headphones, and you're using them for de-clicking. Talking about de-noising, do you find yourself doing this routinely, or is it a skill that you only employ from time to time?

I do have a basic Cedar system, with de-crackle and de-hiss. I used to do a bit of restoration work for a few labels, so I would often get albums from shellac or whatever. I'd be doing a lot of manual de-clicking and then retouch, and obviously whatever Cedar couldn't take out itself; I don't do much of that anymore. I mastered an album recently and de-hissed one track a little bit, but the rest of the album didn't need it. Actually, I used the de-hisser in Samplitude, which isn't bad for the price and suited the track.

Q: Do you do any mid/side work, or do you address stereo width in any way?

Not usually. Sometimes people might ask for it to be a bit wider, but often they find that what I do normally gives it a bit more width anyway, even though I'm not physically widening it. Sometimes I'll use the SADiE "Width" plugin if I need to bring the width in or out. I tend not to do any side-chain stuff at all; I'm a very straightforward engineer, really, probably because from my early days, I was taught to transfer something as well as it could be, and that's been my initial training. Obviously over the years I'd do other things and I'd tweak things, but I generally don't mess around with side-chaining. If somebody in this day and age can't do a sensible mix, lord help us! It's that thing, "Oh, we can leave it to the mastering to sort it out"; I don't think that's right. As a mastering engineer, I'm generally using two channels, but the mixing engineer's got the whole multi-track, endless automation, and goodness knows what else to get it right, plus the budget and the time; then I get given three or four hours to work my magic and sort out the problems that shouldn't be there.

Q: What is your view on the so called Loudness Wars? Is it still happening? Is it over? Is it a non-point? Are people looking at it the wrong way? What do you think?

Well, I think it's still happening, but over the past few years I've tried to not master as loud. Then I do something that I think needs to have a certain dynamic to it. I do that, and send out the reference to listen to, and they go, "It sounds great, but could it be louder?" So you basically just whack it up another couple of dBs, and they say, "Yeah sounds great, thanks!" Maybe that's the way I should have done it in the first place, but I was trying to give them the best sound, not the loudest.

Q: Do you then think, to a certain extent, that laying the blame at the feet of mastering engineers (as some might do) is really just misunderstanding the process in general?

Yes, because you're being told what to do. You're doing what you think you want to do, but then they're coming back to you saying, "It sounds great, but I want it louder." Funnily enough, I mastered an album last week, and I hadn't mastered anything for that guy in six years, and he's in Finland. I had forgotten completely that he likes things quiet. I didn't over compress the project, it was just a nice album, but it was up there. As mastering engineers, we tend to know where our volume control wants to be for the average listening position, so we know how things should sound at that volume. So when I master, I get everything where I like to it to be, and sounding right. Then the client comes back to me and says, "It sounds great, but it's too loud. Can you turn it down about 30%?" So basically, I just turn it down about 8 dB, so it's probably peaking at around −8 or −6, something like that, and he's very happy with it, probably because he likes to listen to things at certain volume settings on his equipment at home. If he suddenly gets something that's up there, it's too much for him; everyone is different.

Q: So with your background in vinyl, when you were working, you had to achieve loudness in different ways, right?

That was normally a bit of mid EQ, generally in the 2–5 k area; also, back in those days, there was a lot of small transistor radio listening and a lot of little labels, Stiff Records, for instance, concentrated on that sound. Dave Robinson of Stiff had a little transistor radio with a particular modified radio compression. I'd cut an acetate, he would take it away and play it through this little radio thing, which would give him an impression of what it might sound like if played on Radio 1 at the time. It was usually that midrange EQ, because the human ears are very conscious of that area, the 2–5 k area. If you've got something going on in there, it will naturally sound louder; Dave and I used to speed up the singles as well, giving a little more adrenalin feel to the track.

Q: You've had such a career; how has mastering changed since you started?

For me, in 1969 I was taught to transfer the analog mix as best as possible to the groove, and we didn't have much in the way of EQ. I cut it as loud as I could without any distortion, without sibilance, and without the groove jumping, which was the criteria then; also, if you were mastering an LP, you had to cut the whole side in one go; you couldn't stop. That was before the days of making production masters, where you'd make a copy, chop it all together, and then just cut it flat. I would make endless written notes of what I wanted to do. I'd start the cut, and it would be adrenaline from beginning to end getting it all correct, making all my EQ, compression, and level changes as the cut went along for 20 minutes or more; plus I'd be making fades at ends of tracks and physically looking after the cutting lathe, making the spirals between tracks, adjusting the groove size, and spacing for specific sections. I'd get to the end of the side, press the lead out button, and breathe a huge sigh of relief if I'd done it to my satisfaction; if not, I'd have to do it all over again.

I did all the cutting for Andrew Lloyd Webber and Tim Rice's *Evita* and *Cats*, and all those huge singles; you probably remember, "Memory" and "Don't Cry For Me Argentina." On those two tracks, there's a lot of dynamics, and I was managing those all the time. I didn't want to take the easy route using a compressor, because it didn't work well on the voice, so I was manually potting everything. I also did some great stuff for the Human League. They used to come down from Sheffield on the train in the morning, master with me in the afternoon, going back to Sheffield again in the evening. They had their own little studio where they recorded their music, and they made quarter-inch manual tape edits. They'd come down with these edited singles from several different mixes, but the mixes didn't flow properly when edited, so there were level and sound changes. In 3½ minutes I might have had 30 or 40 different changes to do all on the fly guided by my notes, super adrenalin cutting those, but so satisfying. These days, you don't even have to worry about it, as it's all on the computer and easy to make those adjustments; you just do them, and possibly more. I'll master a track, and I'll think, "Oh, there's something funny about that one snare drum hit," so I'll take it from the verse or chorus before, as long as it will edit in correctly. I do little things like that just because I feel it will sound better. Carrying on from that, now we have the digital medium, we can go back into everything, we can make different versions without losing anything, we can try edits, and it doesn't matter if it doesn't work, because we can click back and start again. We have automation if we need it. I rarely ever use automation, as I'd rather re-record a section and then electronically edit it on SADiE, as sometimes, a long crossfade between sections sounds better than an automated switch. I do love my L316 multiband compressor, and I love the multiband in the TC 6000, although when I say I'm using them, it's minimal multibanding, usually only half to one dB.

Q: Let's go more into the multiband; what do you typically do with it? Does it serve more as an EQ function, or do you use it to tighten up the bottom?

It's usually a mixture of EQ, overall compression, and multiband compression levelling out the frequencies to how I want them. I have a basic start setup with the Manley Vari-Mu, Prism Masalec, and then the TC 6000 EQ, etc., so by the time it gets to the Waves, the track is sounding how I want it, so the Waves is just there if I need a little more. I've already got the volume where I want it before it even hits the multiband, so I'm only tickling the top end of the dynamics. It's nothing major, really, rarely more than a couple of dBs, but I use several pieces of equipment in minimal ways to achieve the end result I'm happy with, much better than using one processor to do everything.

With the TC, I'm often only adjusting a tenth of a dB. You read in some books that the human ear can't really hear differences less than 2 dBs, which is a load of rubbish. I rarely have anyone attend a mastering session these days, but I did have someone up here recently. It was her first album, and the label's co-owner, whose catalogue I've mastered for the last 27 years, really wanted her to come in, so the artiste and label co-owner were here for the mastering. I was only making 10th of a dB adjustments, and she could hear the difference. That was also due to my ATC monitors, because they are so transparent and portray those small changes. It's very different to a recording or mix engineer, who's often adding several dB's of EQ and heavier compression.

I don't like working with stems and think the engineer should be able to get it right in the first place. I know EQ and compression can upset a mix sometimes, but the mix would need a lot of work to need a stem adjustment.

I've done a few talks on mastering at universities and schools. I do my talk, it gets to question time, most students won't ask a question and just walk off, yet you can see from their faces while you're talking that they want to ask you something. It's amazing, though, because you walk into some of these universities, and their studios are better than some of the professional studios.

The other thing about mastering and cutting engineers is that some of them have to put their mark on it. A few projects I've mastered over the years haven't needed anything. I've played the tape, flat, and it sounds perfect to me, so I've just transferred it that way, and the client thinks it's great; so I got the job, because I didn't feel I had to do anything. With *Evita*, the engineer was David Hamilton-Smith, and I was working at the Master Room. I didn't know at the time, but David had already sent the master tapes to several other cutting rooms for reference acetates, and wasn't happy with the results. I laced the tapes and lined up the test tones properly, and played the mixes, which sounded absolutely fine to me. I thought it didn't need anything at all, so I just cut two flat references, and they came back and said I'd got the job. The other guys had tried to do something to it and ruined the feel David and Andrew had achieved with their mixes. All we did, when we actually cut the masters, was a half-dB lift, halfway through one side on one track, and that was about it. Hats off to David for getting the recording and mix right in the first place.

Another thing are engineers who think they can fix it later in the chain; that's totally the wrong attitude. I think it needs the right instrument, with the right player, in the right part of the studio, with the right microphone; put that on tape and you've hardly got to do anything else. It will sound much better because it will not have been over-processed. It's too easy these days to use loads of outboard gear, when getting it right first is the better approach.

Q: Where do you see mastering heading?

At the moment, I'm not really sure, because I'm semi-retired now, I'm winding down. When I turned 60, I announced I was going to retire, but so many clients I had mastered for, some for 20–40 years or more, said they didn't want to use anyone else, and could I continue mastering for them? So I'm semi-retired, just mastering 2–3 days a week depending on what I've got, and enjoying the outside life a bit more.

Q: If you could go back to the very first day that you stepped into a room to do mastering, what would you tell yourself, knowing what you do now?

I was born to be a mastering engineer. It's never been a struggle for me. It's been completely natural, and I enjoy trying new ways of getting the sound I want. Mastering is something that's inside me; I don't know why people like what I do, but they do, so I have to be doing something right. I've always enjoyed where I've worked. I've loved all of the equipment that was involved in vinyl cutting; I loved all the mechanics of it, too. I'm not particularly technical; I can't build circuit boards and don't understand a lot of the jargon, but I like to know how things work, and why they work in the way they do, because if you do, you probably get that little bit more out of what you're using. When I worked for Nimbus Records, who were a classical record company, in the mid-1970s, I moved to Monmouth, where they had their own pressing plant, recording studio, cutting room, development, and everything else, and they were very, very knowledgeable people. I had the benefit of experimenting with the pressing and galvanics, so I actually made metalwork and worked a press. We did various experiments to improve the processes involved to achieve a better pressing, and that knowledge was invaluable for me to understand the whole process, and that's made me a better mastering engineer.

From my excellent beginnings at IBC Studios, the various companies I've worked for, and now having my own facility, I've enjoyed every step along the way. I couldn't have wished for a better career. I'd tell my younger self to enjoy every moment.

John Dent

John Dent started his career at Trident Studios, working under Ray Staff (also in this book) and Howard Thompson, but quickly developed his own client base and growing reputation with the punk scene in London at the time. His list of clients from the era is a who's who of the music industry, including The Police, The Stranglers, Bob Marley & The Wailers, and Ultravox, to name but a few. John moved from Trident in 1978 and set up The Sound Clinic with Island records and worked with the likes of U2, Bob Marley once more, Fairport Convention, and Peter Gabriel, amongst many others. Later in 1987, alongside Graeme Durham, he started The Exchange in Camden, continuing work with Island and remastering for CD. John then moved to set up LOUD mastering in 1995, where many records were cut for various artists, and shared his knowledge with his business partner Jason Mitchell, joining in 1997. Russ had interviewed John on a few occasions over the decades, and this latest interview was carried out for this book. The news that John had passed away came through just after Christmas in 2017. As you will read in the pages that follow this, his interview shows he was passionate about mastering and high-quality sound throughout his career.

Q: How did you become a mastering engineer?

That's a fairly easy one. I joined Trident Studios after leaving school, and they had a post there for a "tea boy." That was the kind of thing at the time, you had to get into the studio somehow, and being a tea boy, you get familiar with what goes on in the studio. The path I took was tea boy, then librarian. That was literally looking after the entire tape stocks and artist tapes coming in and out of the studio. I did a brief spell tape op-ing (tape operating), but that wasn't very long. Then one of the guys in the cutting room, Howard Thompson, had announced he was thinking of leaving. They were then looking for someone to actually train up and be there as a future replacement, should Howard definitely decide to leave. Out of all the engineers, I was the one that really fancied the idea of getting into disc cutting. I was a big vinyl record collector, I had a pretty photographic memory of the sound of my records. I thought well, although I actually never considered this as a career path, but making tea for everybody, and wandering in and out of the vinyl cutting room did actually fascinate me, because I could see the machinery working, I could see the type of clients that were coming in, and I thought I'd give it a go.

My entry into that world, I actually took to like a duck to water; it seemed to go very well. I understood what Ray [Staff] and Howard were telling me, I was like a fly on the wall for lots of interesting sessions, just sticking around and observing methodologies, approaches, and advice. At some point they let me loose on what they considered to be some sort of minor sessions. One minor session, which was a 7 inch single from The Band Of The Black Watch, "Amazing Grace," and to everyone's astonishment it got to number one. It was one of the first things that I'd ever cut. Obviously, they were impressed, and I went on from there. Howard did eventually leave, and this coincided with punk [music]. As it happened, I had a bit of a flare for working with the punk artists, so my career just carried on going from there.

Q: Was that a personality thing, something about you that resonated with the punk artists, or was it the aesthetic?

I think it was. I was young, probably more of a rebel than I am now; I like the whole punk ethos. Wardour Street and St. Anne's Court, where Trident [Studios] was, was right in the heart of London's punk capital as it was. There were various clubs around, and lots of the people used to hang out in that area. I used to be in a small band when I was at school; I think I just resonated with the simplicity of the message and everything they did. Running in parallel with that, I had started working on lots of reggae. That seemed to be something that was [going] hand in hand with punk music, but also something that I found I took to; I collected lots of reggae records, and I think I had an ear for it. It all seemed to sort of slot together really nicely.

Q: What attracted you to the process of mastering in terms of what you witnessed? Was it just that there was a gig there, and you liked it,

so you settled into it? Something must have kept you interested in that process, aesthetically speaking.

Yes, it did. I think for me, everything seemed a little bit more instantaneous than whole recording process. The recording engineers, the balance engineers, and the tape-ops, they were getting involved in projects that maybe lasted three to four weeks, two months, even. Their kind of involvement could have been no more than 10 to a dozen projects per year, but in the cutting room, you were dealing with possibly cutting eight 7 inch singles a day. With that [kind of] turnaround, and with the speed at which things were being pressed, you'd go home and you'd hear the stuff you'd been working on, either what I'd been working on, or what Ray had been working on—you'd hear it on the radio in three weeks' time. I actually found that quite exciting; it was making music and getting an instant reaction. I kind of enjoyed it and never really felt any need to change after that. As time went on, my client base seemed to get more and more important and bigger. There were certain key recordings that I mastered that really set the world alight, in a way. In 1975, I mastered and cut the Bob Marley *Live! At The Lyceum* album; "No Woman No Cry" was on that album. That was the point in time where Bob Marley really became a worldwide, household name. Having had similar experiences with other artists, I found it exciting. It was the excitement and the instantaneous quality of it.

Q: What has changed about the process of mastering, with regards to getting this instant feedback from the industry? Are the turnover rates the same?

For me, they're not quite the same. I actually now, and have done for quite a few years, been involved in the more lengthy projects. Back in the day, certainly across the punk period, the 7 inch single was really king. Everyone that wanted to make a noise knocked out a couple of tunes, it became the 7 inch single. There was a lot of stuff being produced at that time, huge amounts, because it was very accessible to everybody. I think as time has gone on, the work I do now is linked to artists that perhaps have struggled to find an engineer they relate to elsewhere. A lot of my approach actually hasn't really changed since then. I still feel like I hear the same way, but that causes me a lot of problems, because I grew up and cut my teeth listening to high-resolution sound on tape. That high-resolution sound to me was the thing that actually [gave] a sense of reality in the recordings. Nowadays, I consider a lot of the stuff I get given to work on [to be] considerably poorer in sound quality. I struggle to listen to 44.1 and 48 kHz recordings, because I find them so empty and lifeless. It's to do with how I respond to sound and recordings.

I think for me, the bigger projects I get are sometimes more complex. I'll give you an example of that. A couple of years ago I was asked to work with Goldfrapp again, and they were quite interested in sending the tracks onto the vinyl lathe, then back off on a good deck, then using that source material off the acetate, to actually master the album. That was really interesting; they were drip-feeding me tracks one at a time. I would master it via the lathe, via the cutting system. I would also do it a number of other ways, as a comparison. So I was giving them maybe three or four different versions for them to actually choose how they wanted the album to be put together. In the end, that Goldfrapp was 7 out of 10 tracks through the vinyl method, and I found that very interesting. The whole thing was unattended, with a panel with at least five people approving. Everything that I did got approved universally, all the way around. I was like in my own mastering environment, doing all this stuff. The upshot of that is, for that year that album was voted by the Music Producer's Guild as the best album of 2014.

We auditioned that album; I was one of the judges—I wasn't allowed to vote on it, but I was one of the judges. Out of all the tracks in the room that people were listening to, that album was actually listened to twice; people actually asked to hear it again. That's kind of the current John Dent, the totally self-indulgent guy, and if people have the budget and are prepared to indulge, then I will doubly indulge, as it were. My throughput now is more like that, rather than just bashing out 7 inch singles. It's a different world, you know.

Q: In following up on your thoughts about having a hard time listening to 44.1 k/16 bit, in terms of them sounding lifeless, do you think that there's anything to say about the change in production circumstances themselves? Nowadays, you're dealing with people who are often working on laptops, or aren't going to book out a space for a period of time and develop the album.

I think there's been a huge shift away from how I feel things ought to be recorded; people don't have the budget. I can understand the financial side of it, I can understand the convenience side of it, because people want to create on the train, on the plane, and in their homes. I can understand all of that, but I think some really important elements in sound quality have actually gotten lost. It's a generation thing. Lots of people now have been listening to CD and mp3 since the '80s. A lot of people just think that's all sound is; they are totally unaware that before that, there was this other world of sound. They will only hear that other world of sound through the CD and mp3 mediums, because they would go out and buy reissues. If you were to take the time to go out and buy the original analog vinyl records and A/B them, which is what I do here. . . . I have a Beatles single record that I actually sit down and A/B with my clients, before I master their CD; I actually show them how bad CD is as a medium. And then explain that we're going to master it differently, and it's going to be somewhere in the middle. I hear this all the time. A good analog vinyl is about five times the resolution of CD. It's like looking at a photograph, a 12 megapixel photograph, and a photograph at 3 megapixels—there is a difference, and I instantly spot the difference in audio.

I've got the "Something" single by the Beatles I use for this. It was cut to vinyl in 1968, and I've got the Abbey Road official George Martin CD version. I start playing that to someone, and they hear it, and they sort of sing along to it. I then switch to the vinyl [version], and their jaws drop; it's jaw dropping how bad the CD is. The CD just doesn't have the same quality. It's like George Harrison sounds like George Harrison on the vinyl. It sounds dreadful on the CD, and I'm sure it's the same tapes.

You don't have to do it with that record. I have endless amounts of vinyl here which I can do the same comparison on, because it's a resolution thing. In my mastering room, I've actually deliberately set it up to be very high resolution. It's very obvious, sitting here, how good a lot of these analog recordings actually are and how awful the CD medium is; it's soulless.

If you listen to Led Zeppelin off the record, and you play the CDs, and you think, "What on earth has happened here?" It's right across the board. It's a global thing, where people are just kind of listening through a filter. In the old days, the filter was the cartridge or the stylus, on the record player that you're playing. What happened in the old days was that you bought a record and listened to it on whatever you had. But if you went out and invested in a good record player, the information was actually on the record, so you actually heard more and more—that's what hi-fi was all about. Nowadays, the whole world is listening through the filter of AAC, or mp3, or CD files. To me, it's too drastic a filter, a much too drastic filter, so that's how I view it, and I can mathematically prove it.

Q: What's interesting in terms of your approach to mastering is that actually when you master, you master with vinyl in mind, whether it be for anything else or not. Many of the other engineers have said that they master with the digital file in mind and then cut that to vinyl or other distribution types. Everyone's saying the same sort of thing, but in very different ways. How have you set your room up to achieve these different formats? Is it different from anyone else's?

It probably is different. I've worked with a mixture of analog and digital gear. I'm always encouraging my clients to work at higher sample frequencies, and actually take the time, just about every session, to actually explain to people that you can't just record it at 48 k and then up-sample it somewhere in the chain: you have to actually go out and do snapshots of recordings at a much higher sampling rate. Clearly people have technical problems with their computers coping with this. It's only been in the past few years when powerful processors have allowed larger numbers of tracks all running at 96 k, but we're obviously heading in a better direction from that point of view.

I've got clients where they've made the change, and they would never go back. I've got a client in Columbia, and I got him to switch to 88.2 k, and he records 30-piece salsa bands, and whatever. His recordings are fantastic, absolutely fantastic. They're not analog, but they're certainly nowhere near the plainness of the CD. What I do is, I'll keep everything at a higher resolution whilst I'm mastering, and then at the end of the chain I'll throw two-thirds of it away and call it a CD; that's kind of how it works. I do encourage clients to actually release the higher resolution versions, and we use those higher resolutions to cut the vinyl record as well.

Q: In relation to your LOUD Mastering company, and I don't want to get too much into the "loudness wars" thing, but I was wondering if there was a comment being made there?

I think the first thing is that I used to wear very bright shirts. I did a bit of body boarding, and I'm known for my "loud" shirts; that's where the name came from. Also as a vinyl engineer going way back into the '70s, it was the thing on everyone's lips. People came in and said, "Come on, John, make sure it's a nice loud cut, we need to get this on the radio." It was just something that was in the air all the time. It was in fact something that you had to take note of, because if you didn't get a decent level on the vinyl single, there were other engineers around in other studios that did. You had to work out how to use a limiter properly, and what sounds restrained. Loudness is a subjective thing. It's not just where meters peak—it has to do with how your ear and brain perceive the sound. I think that taught me a lot about subjective loudness, so what we should be called is "subjective loud mastering," as opposed to "loud mastering," because it's not just about putting things through limiters; it's about soundstage, I would say.

Q: Will you speak a little bit more about alternatives to reducing dynamic range to achieve loudness?

I think what happens is, I have collections of music here that have a big sound in the speakers but aren't necessarily totally hammered from a sort of "naff" limiting point of view. Sometimes these sounds have to be almost recorded and engineered in right from the very beginning. Conceptual things:, how much space there is around the drum kit? How much reverb? How the snare sounds? I'll give you an example: I've got some hip-hop stuff on my computer where if you look at a snare hit on a spectrum analyser, just about every frequency on this analyser lights up, from 20 Hz to 20 kHz. Bands come in with a snare that's just like a single frequency hit with no width to it. That won't have subjectively a big sound, because it's just sort of a very narrow sounding snare. I'm always encouraging people to rethink the way they work and to try and engineer some of this bigger, wider sound rather than relying on the L2 [limiter] all the time. It's just being aware that you can always move in this direction. When I use an equalizer, for instance, I'm maybe using it in a way that other people would find a bit strange, but what I'm listening to is the size of the sound, not trying to over-boost anything that doesn't need over-boosting. I've had a particular passive equalizer, Vic Keary built me one, which is actually built to my own design specifications. He had the designs originally, but I said I needed this, this, and this. I've got his standard front panel, but as the knobs turn around, they go off the scale, and there's a load of hidden settings on it. Things like air, sitting at the very high top end. I can just dial it in and have a listen. Sometimes, that's all you need—take a closed-in recording, and whack in some of this EQ for air, and all of a sudden it sounds bigger, and you haven't done anything to it other than put the air in. It's listening from that point of view. I don't have a standard way of working, so what I tend to do is listen to a project first before I touch it, and discuss with the client what it is they actually want, and what they're trying to achieve. Then I just look at how the sound moves towards where they want it. It's not necessarily how I want it, but quite often people just want a very simple thing: they just want a slightly bigger, louder version of whatever they worked hard on to produce. I can actually do that quite successfully. I've had quite well-known engineers in here where they've heard the final mastered thing, and it's been technically about 8–9 dB louder than the file that they gave me. They say, "Well, John, how did you do that, because it doesn't sound like it's changed much?" And I don't tell them, it's just the way I listen. I use my gear to achieve that end. I accept the fact that there may have to be a little bit of compression or limiting, but you don't have to just hit the compressor straightaway; you can engineer it in other ways, so that's what I do.

Q: I'd love to hear a little bit more about the fact that loudness was always a concern, but it's a very different loudness than what's been so controversial recently with the "loudness wars." For instance, a lot of historians of music will point to the Beatles records, or the Motown records: these were known to be loud, and that was a good thing. Do you think that your ability to achieve loud masters has to do with your experience working with vinyl, where you have particular limitations that you maybe don't have in the digital medium?

It definitely comes from there, because the whole thing was [that] you want a decent vinyl engineer, unless you could achieve a range of dynamics from what the client wanted on the record. The loudness thing was definitely for a lot of people, something they wanted, and as a vinyl engineer you had to know how to do it. It was particularly tricky, because the louder you cut a record, the less record players could play it back, the more sibilant distortion you got on cheaper pickups. There was definitely a way of shoehorning a bit more level onto records by trial and error. You got used to working a certain way. You got used to the differences between valve equipment and transistor or IC [integrated circuit] amplifiers. You got used to what happens with phase relationships between stereo, and things like that. You just got used to all this stuff. It's something ingrained within me, in terms of my programming, as it were, that still exists.

I think the modern "loudness wars" is actually something totally different. I think that's definitely driven by the very poor reproduction systems that a lot of people are now listening on. If you do the mathematics for a CD, for instance, and I think it is for every 3 dB increase in level, you need to double the power of your monitoring system. I did work it out once, but I think for a CD you need something like a 500-watt amp and 1,000-watt speakers to hear the proper dynamic range of a CD, as it is. When someone's listening on a 0.5-watt speaker on a laptop or on a phone, the only thing that you can do to really make them hear [it] is to compress or limit the hell out of the source material. To me that's what's driven this whole "loudness wars" thing. I think that if everybody was listening on hi-fi systems, it wouldn't happen, but that's not the way the world is consuming music now. I have this debate with clients every session, because the client quite often has to understand that the dynamics that they've engineered into their recording may only be audible on certain systems. Time after time, again, I sensitively master something so that it's not overcompressed and peaks in the right places, only to find the client ringing me up and going, "We can't hear it on our laptop?" And that's true. You have to, and I try and pre-empt this in the kinds of discussions and tests that I do for people, but I think the "loudness wars" really, now, is driven by these very, very poor reproducing systems.

Q: In terms of your process, you listen to the material and then begin with a consultation where you manage client expectations?

Yeah, it has to. I'm one of these people that doesn't have any set system. I don't just cram the sound in and do it. I hear of stories about mastering studios elsewhere, where the client has no say whatsoever, you know. They just put the track up, shove it through a few plugins, shove it through the desk and go, "Right, your track's mastered." Well. that's not how I work. What I'm trying to find out usually is an intended concept for everyone's piece of music, because by the time they get to the mastering, they've lived with it for months, and I've just heard it. So, I don't have any real right to imprint my style onto it. What I need to do is find out what it is that they actually want from their music. We sit down, and I listen to all the tracks, listen to it and decide which tracks perhaps are the biggest, fullest tracks, and go from there, really. We look at a method of applying the mastering. Some people have very fixed ideas, and some people are open to experimentation. Sometimes after a bit of experimentation, there's this kind of glow in the room, where you see a whole band's face light up, they're hearing something that they respond to, and I respond to it as well, and think, "Well ok, there's something in this." We just find a template, and find a style of working, and then go for it.

Q: That ties in with the collaborative theme we're seeing in these interviews—what you're describing is a very collaborative process, being part of the production process, you're speaking with the people involved, and making decisions with them, and that seems to be one approach to mastering. Then there's another approach to mastering, where it's almost like an "arm's length peer review," if you know what I mean?

The online mastering thing, the "black hole" that you feed your track into and it comes out LP'd out the other end, is a good example of that, where no one's asking anything; they're just literally doing what they think is mastering, and the client ends up being stuck with it.

Q: Does it have to do with, again, your experience working with vinyl, where sometimes there are certain balancing moves that can occur in mixing, that just won't translate? If you're not aware of the differences in the different formats, that you wouldn't fully understand that, so you have to communicate in that respect.

I suppose that is part of it. I find this, "You can't cut anything to vinyl" thing a bit of a myth, because I can actually cut quite a lot of stuff to vinyl quite successfully that other engineers don't seem to be able to do. A lot of cutting rooms, for instance, will mono the bass. I only do it reluctantly if I've got a very long album needing fine grooves. I've rebuilt the groove packing—the Vari-Groove system that I have here, I rebuilt it from scratch. Mine seems to sort of get stuff on there without touching the stereo of the sound, or without doing too much. There's a lot of myth and legend about disc cutting, but it doesn't have to be that way, actually.

Q: Is mono-ing the bass not as necessary as many say, or is it even a myth? Are there others?

It's not a myth. Sometimes it's a functional thing for artistic reasons. Sometimes the bottom end sounds actually a bit more powerful if both speakers are moving in unison, on a stereo system. You see websites offering to cut vinyl. These websites actually explain, "Oh yes, while we're cutting, we will be mono-ing the bass, and we will be shaving off high frequencies." Some of these websites actually list what they're going to do, or they'll make suggestions about what they might do. I go and look at all this, and I kind of know what they're doing, because that's, to some degree, how I was shown originally. However, in time, I've realized that a lot of that is actually unnecessary. I will do little nips and tucks if there's one particular track on an album where the stereo toms are going berserk and you just can't cut it. What I tend to do is look at that track in insolation rather than blanket mono the entire bass [end] of the whole album. That kind of thing is how I would work.

Q: In general, how do you see the mastering process to have changed with regards to its position within production?

I don't think it has changed. I think mastering studios nowadays have been given a gift. That gift is the way people work in digital workstations. What happens is that day after day I'm getting files in. They're peaking at zero [0 dB]. They're effectively up at maximum, and you play them, and they're subjectively, say, 8–10 dB quiet for general consumption. I think the way people are working now, a lot of people working on their laptops and whatever, they don't actually know how to produce a better-level file and a more "closer to mastered" level file. I'm not talking about just patching in a couple of limiters as plugins; I think that there's a whole work ethic that people have lost the plot with.

   I've got an engineer friend who used to work for Chris Blackwell at Basing Street Studios, and when he gives me files, they sound absolutely fantastic, because he's an engineer who used to working with tapes. He actually knows how to craft the sound so that when you get a sound to a mastering studio, it may only need 2 or 3 dB of tweaking, but the basic essence of what's there, he's sorted it out. A lot of people actually don't know how to do this anymore, so that continues. People go to colleges, they study at all these academies, and they still don't know how to do it. There is a need for that independent set of ears in a good mastering studio just to kind of tidy up what they give you and spit it out at a more acceptable level. A lot of the skills base in recording had got lost.

Q: Is this through education, say at higher education changing the landscape, or responding to it?

I've got clients who I've been effectively mentoring from their first projects, who are now very switched on and can actually give me very good sounding files. They've cottoned onto what ought to be done. It's hard work, because I think that there's this other weird thing going on, where students are being taught that you can almost make things do anything. I remember when I was lecturing at Red Bull Academy, and people were asking me, "Well, how do I make that saxophone sound like a '50s saxophone?" I said, "You record it properly." Take the '50s saxophone, and you don't record it on your laptop, you record it on a decent tape machine. There's all these things—they have no idea that the actual method you use to source the material is really vital to get the final sound. They think it's all tweaking, and plugins, and fiddling, and it's not; it's a totally different mindset to understand.

Q: People think they can "fix it in the mix," but really nothing is more important than getting good musicians, performing with a great arrangement, and recording them properly.

That's right. It's to do with the whole mindset behind it. I don't think modern technology really helps with that. If people had to go back to a multi-track where you couldn't just cut and paste, you just had to get it right the first time, they'd suddenly realize that it is all about performance, and the difference in the mics you use. It's all got a little bit twisted and lost over the years.

Q: Not necessarily in terms of a list of equipment, but it seems that every engineer's set-up is different, so everyone's building their own "instrument" as it were, and we'd love to hear about anything that informs your technique or aesthetic.

The room I've got here now is kind of end result of a lot of fine-tuning and a lot of working. I built my monitoring system to listen to vinyl records, the detail, the general kind of vibe that the vinyl record gives you, mainly because I do consider them a much higher resolution than a lot of the work I'm working with. I built my system as a vinyl record player, if you'd like. I then just feed in the signal chain that I'm working on, as a monitor, into that. It's very obvious for instance, if you're using the wrong cables, or you're using the wrong amplifiers or the wrong player on the Macintosh. You can play the same file on a dozen different players on the Mac, and it sounds different on every player, so you need to make judgements. Sometimes the judgements are related to overall sound quality; sometimes they're related to particular tracks that you're playing at the time. I suppose what I'm doing is I'm giving myself the hard task to reference everything back to material that I know will sound really good on the record. I just use that as my daily process. All my decisions about which bit of equipment I have and how I use it tend to be related to what you hear by comparison.

I don't have any particular rulebook. I do use short cables. I don't have a patch bay. There's a very good reason why I don't have a patch bay. When I worked for Island Records, I had a studio called the Sound Clinic. When I first put that studio together, it was literally piles of equipment where I worked, and all linked by XLRs in the back; I just got on with it. I was cutting a lot of hit records. I had clients like Peter Gabriel, Dire Straits, and all sorts of people. At some point, I persuaded Island to invest some money so I could tidy my studio up, and I spent about a week rewiring all the equipment into a patch bay. I then tried to do my first session and realized that what I'd done had actually ruined what I had. I tried the second session, and I thought, "Shit, I've actually completely ruined the sound that I had that was very successful." So, I literally dismantled it all and put it all back as I originally had it, and then all the sound came back. So I learned a very big lesson there, that you don't work with a patch bay.

I've adopted a lot. I still haven't learned about everything. My equipment here is sitting on rubber mounts. The chassis on a lot of my equipment don't touch [each other or the rack]. I can choose how to ground that particular piece of equipment. I'll listen to the noise floor, and I'll try a different grounding, and decide which noise floor I particularly want. Things like that are inbuilt here. The other thing I've got here is that I've got balanced mains, so there is no 50 cycles of hum on anything. That makes a huge difference to how you work. There's a lot of stuff here that I've kind of put in and left. Once you collectively have all that, then you've got a nice mastering studio.

Q: You've mentioned that you don't have a particular workflow or set routine, but are there particular things you do regardless of the project or the programme material?

That's a good question. I suppose from the monitoring point of view, I've got a set of VUs here, which are linked to the output of my Sonic Solutions computer. I do actually study the VUs—that's a disc-cutting thing. I will look at the way the sound is on the VUs, and see where things peak to. It's quite interesting, because all CDs, I would say, within reason, are peaking at CD zero, and they all have different sounds within that. The sounds that people generally like and want are more visually obvious on the VUs. VUs behave in a certain way, and I definitely get that from disc cutting. And it's something that I kind of got used to over the years. So that's a common thing: I will always study what's going on on the VU.

Q: What is it about the VU that you think is more revealing in that way?

The VU definitely relates to your ears in terms of perceived volume. A PPM does not give you the right feel of volume. You can have a file peaking at PPM zero that sounds 10 dB quieter. If you look at it on the VU, the VU is hardly moving. So that tells you that it's quieter, and the VU is telling you what your ears are telling you. If you then find a way of amplifying it, and suddenly the VUs are now at −2 or −1 dB, you know that the client is going to hear a fuller, louder version of what you've been working with; it's a very good tool. I don't really have any other set way of doing things. I have a number of players on my computer here. I've got both half-inch and quarter-inch tape machines. I quite often connect my equipment in different running orders, depending on what I'm doing; I always keep a bit of an open mind when I'm working. There are some clients that just want something very simple, so I'll explore whether it can be done completely digitally. I've got different ways of doing that. I've got nice clocks and can deal with the clocking differently depending on what's needed. So I don't have any real rule. I suppose my sessions actually take me longer than most people—I'm not the greatest economic production line here. What I have got is a very large trail of very satisfied clients behind me.

Q: And expert ears, right?

Yeah, it's all that experience and all that technical stuff, all rolled into one. It's the John Dent experience, you know.

Q: What is mastering, in general? What is expected of you, if you had to put it down to something?

There are a couple of things. You've got a world of music out there, it's established. People know what's what. It's an absolutely massive body of work out there of all different genres and all different styles. The new work, the stuff that comes in, may have the kind of ideas behind it, but it may not quite fit with that other body of work that's already out there. It may be a simple level thing, it may just be a tonality thing. If all the stuff out there has a fairly warm, big, full sound, and someone gives me a very thin-sounding, harsh, digital file, you have to ask the question, "Is that how they want it, or is that a mistake?" I think my view of mastering is that you're just making that bridge between the way stuff has been brought in and presented to you and the wider world that's out there. If you make that bridge correctly, then that record will then be listened to, people will understand what it's all about, and with a bit of luck, it will be successful. It's sensitively and creatively making that jump, between source and the real world. Some recordists can actually more or less do that themselves, with their recording style. There's such a big difference in the way people can create masters—a huge difference.

I remember having a very sort of drunken discussion about what I do in the pub once. Someone said, "What? You mean you charge them money to tell them there's nothing wrong with their tape." And I said, "Well, yeah." Although you're charging them because you're using your time to evaluate it, and coming to the conclusion that actually they've done it right, but that may only be one in 50.

Q: So you need to have that expertise to be able to make that call, and judgement, in any event?

You do; it's like any creative medium. I always relate what I do to a professional analog photographer, who will understand how to compose and take high-resolution photographs. That photographer will be up against the whole world of phone and digital cameras, where people just take snapshots: there's a difference. That's what I do.

What I do: it's advice, it's suggestions, it's encouragement. It's all those things. To kind of help people decide how they want their stuff. People sometimes don't know what they want, and some people have to be shown something. It's a whole mixture of things. The real world is out there, it exists. You have to be aware that stuff won't stick unless you make it, sometimes. So that's what I do.

Q: We're finding that there's a bit of variance in terms of mastering work, but also the industry itself is struggling as a whole, and many of the sub-industries are also struggling. We are finding that somehow mastering, in some circles, is adapting by becoming almost an added step in mixing. Do you have any general thoughts about where mastering is heading in the industry and as an art form, in the face of what's happening in the recording industry?

There's two issues there. What's happening with the record industry in general, I think, is absolutely shocking. I think to be handing half the income of the industry to computer companies like Apple and distribution companies like Amazon is absolutely crazy, because the industry is struggling, and artists themselves are getting insignificant royalties from places like Spotify. The whole record industry has been hijacked by the computer industry very successfully, so there's no investing in the industry. That's my opinion on that side of it, and I totally blame the record companies, because instead of being a unified force and sorting out their own distribution system, they're quite happy to hand half their income to the likes of Apple, and I find that astonishing.

Where mastering is going, I think it sort of depends on the individuals involved in the recording chain. There is pressure on mastering at the record label end to cut costs and not even bother. I think a lot of that has come about through engineers: when they do a mix, they may have mixed their stuff peaking at zero, but subjectively it's 8 dB quieter. What every engineer that I know does is always bump it up through a limiter, to hand it to a record company for the A&R department to actually hear what's going on. You do get this phenomenon where they've been listening to this very rapidly compressed set of files, and then they send it off to a mastering room, and actually it doesn't quite come back the same. They actually prefer what they got from the studio. That's a bit of an odd one, because it almost means you have to try the mastering in order to decide that you don't want to use it. Everyone has these tools now, so what we have to do is actually think outside the box and decide, "Well, is that the right approach for it?" So, I think on a sort of consultation level, and on an experience level, and perhaps having a broader, wider view of things, then mastering is still relevant. I think there are definite pressures within the mastering industry that makes record companies themselves question whether to spend time with a mastering engineer. I've noticed it a bit, because I've been in the industry so long, a lot of the client base has changed. People that knew about mastering on the A&R side of it, they've all left the record companies, and they've been replaced by much younger people who don't actually understand the significance of the mastering process. It's a kind of "era" change been going on, and it happens all the time. I'm sure all established mastering engineers would say this, that they had a whole heap of work when so and so was working at Virgin, and then now they've left, the work's dried up. It's a constantly changing industry. Having said that, there are a lot of people out there with inquiring ears, and young musicians who really do feel that they want to explore what can be achieved in the mastering process rather than them struggle to do it themselves. It's a personal thing, as to how people view what they want from the sound.

Barry Grint

Barry started his career at Trident Studios in London as a tea boy. Within six months, he was running the tape copying room and mastering records within a year. He became chief mastering engineer and retained that position when the studios were bought and renamed Audio One. During this time, he mastered vinyl for Warner for the UK and Europe, with acts including Madonna, Prince, and ZZ Top. He also mastered Simply Red—*Holding Back The Years* and Aha!—*Take On Me* as well as other projects.

He moved to Tape One Studios in London. This facility was at the forefront of digital technology, with digital transfer consoles and the first digital disc mastering desk. Projects included The Shamen—*Move any Mountain* and The Farm—*All Together Now*. When the studio closed, Barry was recruited by Abbey Road Studios as an additional engineer, the first to be taken on for 10 years. Work here included Oasis—*Definitely Maybe, Whatever, Roll With It*, and *Some Might Say*.

In 1998 Barry founded Alchemy mastering currently located in Brook Green, Hammersmith, London. Together with Matt Colton, Alchemy has been at the forefront of half-speed vinyl mastering. A member of the

Music Producers Guild Mastering Group, Barry instigated the embedding of ISRC data within broadcast .wav files. He worked with the European Broadcasting Union to define the standard and with all of the major mastering DAW manufacturers to implement the standard and to ensure cross-platform compatibility.

More information can be found via the alchemy website www.alchemy mastering.com or at Wikipedia https://en.wikipedia.org/wiki/Barry_Grint

Q: How did you become a mastering engineer?

I'd always wanted to work in a recording studio. I had started doing Hospital Radio and kind of figured out for myself how to use tape machines and how to create tape echo effects, that sort of thing. I then took on a guy who wanted to do Hospital Radio, as well. He then managed to get a job as a runner at Trident studios, and when they were looking for someone else, he got me in. We were working 24 hours on, 24 hours off, rolling seven days a week. The house engineer was "Flood," who was working crazy hours, and I came to realize that I couldn't hack listening to the same track over and over again for 3–4 days. I discovered Ray [Staff] in the mastering room, working a day on and a day off with Mike Walker. I hadn't realized that mastering existed up until that point, but then it just seemed to be the route to go down. So instead of becoming a tape operator/assistant engineer, I went into the tape copy room, and then into cutting vinyl.

Q: How did you learn to master? You found the room, and once you were in it, how did you learn the skillset?

Trident had a contract with WEA to master the vinyl for all of their international acts, so it was all unattended work, but there was a lot of it. Then, Ray decided to put back into service a moth-balled lathe, and Sean Davies built a room around it to get it going. I would then receive quarter-inch EQ'd production masters from the cutting rooms in the US and cut the vinyl for UK and Europe, which at that time we were doing in the region of five sets of lacquers for each release. They would then go off to five different factories to get the stock out fast enough. WEA were also using a copy room that wasn't particularly good, and so the masters were supposed to be 15 ips NAB Dolby A, but sometimes they would be double Dolby on the left and no Dolby on the right, or only Dolby on the right, all of that sort of thing, which you come across when you're working in the tape copy room making the production masters for other territories. You learn that sort of thing, and then you learn what the mastering engineers have done, what will cut, what won't cut, but without having to go from scratch. The system that I was using was made by Ortofon, which gave a better finished result but was more fragile than the Neumann system. It was a question of learning how to get what you want onto the vinyl but without using brute force.

Q: Can you elaborate a little bit on the differences?

The Neumann system uses two solenoids mounted at 45 degrees onto a torsion bar, which then has the stylus put into the bottom of it, and it's the action of the pistons, either in opposition or unison, moving the torsion bar that gives you the groove shape. With the Ortofon system, the two pistons are pointing straight down, and there is a bridge between them. The stylus is then mounted in the middle of the bridge piece, and again, you have the pistons acting in unison or opposition, doing the same thing. However, the whole arrangement was more delicate; it was designed more for cutting classical music. Back then, we thought we were cutting loud, but it's nothing like the level we're doing these days.

Q: I know this is a bit of a broad question, but we'd like to get your specific take on it. If you had to explain it to yourself when you were first entering the world of record production, what is mastering?

I think there is no single answer, because it depends on what's come through the door. If I had an EQ'd master coming through the studio and was asked to cut vinyl, because the client decided to put it out on vinyl, I would say that was mastering. If somebody is putting together a compilation CD, piecing that all together, and getting tracks that have been mastered for release in different places, to sit together, I'd say that was mastering as well. The obvious answer as a mastering engineer, when a track comes in from having been recorded and mixed, is that you're doing the final adjustments, to get it to sound the best that it possibly can.

Q: Would you say that there's a bit of an element of curation, in the sense that, it is expected that you as a mastering engineer will ultimately know what the best sort of balance for a mix is, and your job is to get it in that shape in some cases?

I think the position of the mastering engineer these days is almost unique, in that we work in rooms that are acoustically treated. We only ever work in one room, and so you become very familiar with how you expect things to sound in your setup. The advent of computer-based recording has created the opportunity to record in places that people could have only dreamt of being able to do in the past. However, it does mean that you're going somewhere because you want to capture the acoustics in that place, but you don't have an acoustically treated control room handily located beside it. Also, engineers and producers tend to have to move from studio to studio; therefore, I think that what they're able to produce is quite clever, because they're never ever really given a chance to settle into an environment and get to know it intimately.

Q: There tends to be a reputation of the mastering engineer as almost like a Tolkien-esque wizard, and that they can do things to mixes to improve them, that no matter how long you've been in the industry, you're not quite sure what they're doing, but you know the results are amazing. Was that the case when you first started, when you were cutting vinyl?

Not when I first started, because my brief was to take what was on tape and to transfer it to vinyl unadulterated. It was only once I had cut my teeth and started doing work that hadn't been mastered by someone else that I was then expected to contribute. I think the mystery extends from the vinyl era, and to a certain extent, that's still the case now. Lots of people are mastering engineers, but if you're somebody that has a lathe in your room and you can use it, that kind of separates the men from the boys. I don't mean that in an arrogant way at all: it's just that you can't simply walk up, push the buttons, and expect to get a result; you have to have gone through a learning curve to be able to get a result from it [a lathe].

Q: Let's talk about automated mastering services. What is your opinion about these services, and what do you think that says about what people understand about mastering in general, that they think they can upload a track, pay £5 or $5, and have a "mastered" track back? I understand it satisfies some aspect, but what do you think about that?

I think that there are fewer people who actually understand what contribution mastering makes in the whole process, and to a certain extent, it has become a box-ticking exercise. There also seems to be, particularly within the majors, quite a high turnover of staff in their production departments and in A&R coordination. Also, they are so busy that they never really have time to get a handle on what it is that we're doing as mastering engineers. As long as they've sent us something, we've sent it back and it is approved, then it's been mastered, therefore it's okay!

Q: What are things you typically listen for, and what aesthetic areas might you address in a mix, regardless of genre?

The way I interpret the role of the mastering engineer is as an educated listener, in that we don't know all of the hassle that's been gone through by the people involved to get the track to the state that they're giving it to us in. We don't know that the producer has insisted on playing guitar, which none of the band wanted, or that the drummer has punched the vocalist, or anything like that. We just get the track. We don't know about any of the history. We just sit there and ask ourselves, "How does it sound?" It's really as simple as that. If it doesn't sound right, and it can sound wrong in any number of ways, hopefully you've got the expertise to be able to change it into how you think it should be. That's coming back to the notion that "I only work in my room, I know how I expect things to sound."

Q: Do you notice any sort of commonalities, any trends, in the pre-masters you're receiving now, that were maybe different from when you first started?

An awful lot come into the studio pre-squashed, which leaves you virtually nothing to work with, and that used to really annoy me. Then, I was talking to some remixers and asked, "Why is it that you always squash things before sending them out?" Their response was that they couldn't know for certain whether the record company will actually get their mix mastered. What had happened to them in the past was that their remix was one of several on a release, and their track sounded really quiet because they had left room for the mastering engineer to work, and so their mix didn't come across as well. It's a question of the record companies being clear with the mix engineers whether or not they are aiming to get the release mastered, and if they're not, well, then the mix engineer, just for their own reputation, will have to do a bit of squashing.

Q: Do you have a preferred signal chain, and if so, will you take us through it?

No, I don't really have one. I think you just have to take each track as it comes, and get a feel for the best way to achieve what you're looking to achieve. Although, even then, sometimes after having worked on something for a while in a particular way, accept that you're actually going down the wrong path, and then have to go back and try it a different way. It may be that you decide you're going to do it all analog, you may decide you're going to do it all in the box, you may decide you're going to do a mix of both, but you can get it wrong. Then you just have to confess, back up, and go for it again.

Q: So the most important piece of gear, then, is your ears?

Yes, at the end of the day, and to have kit that enables you to get to where you want to be, which sounds a bit odd, but you can have a piece of kit with humongous amounts of buttons, levers, and who knows what else, but if it takes you twice as long to get to the results you're after, as compared to a more straightforward, good piece of kit, then I'd go for that more straightforward, good piece of kit.

Q: You've mentioned that vinyl cutting is one of the things that separates the boys from the men, and probably one of the things that accounts for mastering being considered by many as a "dark art" or a "Jedi art." De-noising tends to be another one of these things, and I'm wondering why, because you hear de-noising bantered about as one of the things mastering engineers do at mix level, and we're trying to see whether or not this is reflected in the actual practice of actual mastering engineers. How often do you find yourself doing straightforward de-noising, and can you give some examples of things you might find yourself routinely doing?

Well, I wouldn't routinely de-noise; the whole thing with mastering is that you do what needs to be done, and what doesn't need to be done you leave well alone. There's an artist called Bert Jansch who was part of Pentangle; I have just received all of the original quarter-inch masters and was digitizing to 24 bit/192 kHz, then going through to do whatever needed to be done, so that they could be released onto digital CD and vinyl. Within that process, there were clicks and noise to remove, but not really much processing in terms of EQ and compression, because what you're trying to do is to present to people what was recorded. It was recorded very simply using two microphones, or one microphone with the early material, because it was just mono, and then straight down to tape; what went onto tape was the master. For those projects, the point of the de-noising was to change the relationship between the audio and the noise. If the noise detracted from the audio, then to try and change that balance, but not necessarily to remove that noise completely, because you have to be aware of the artefacts that happen with de-noising. If you can just get it so that the noise isn't intrusive, then that's the result to go for.

Q: Given your background with vinyl in particular, besides compression and brickwall limiting, because they've been the "hot button" issue with mastering for so long, what are alternative ways of achieving loudness?

It depends how you're measuring it. You can have loudness that bends the needles back, and you can have loudness as you perceive it to be, and I've heard that achieved by using echo, EQ, or parallel compression—there are a number of different ways to get there, and again, it depends what you're working on, and what's the best way to go about it.

Q: I'm intrigued to hear about the use of echo; will you expand on that a bit?

It's not something that I've had experience of doing, but it's certainly something that I've had experience of hearing, particularly records that were released during the '60s. Some of the pirate radio broadcasts used echo on the DJ's microphone in order to make them sound more powerful, so it is a tool that can be used.

Q: Can we talk a little bit about parallel compression? What is your experience with it?

It was something I had started to experiment with after having found it being used in tracks that were coming over from America. The idea is that you're adding more power to a track, but you don't end up with something that has no peaks. It's adding a powerful core within the original track, and as with anything, there are artefacts; you have to be really careful about unusual pumping effects, not the typical ones you would expect with compression. Sometimes you have to ride the gain of the parallel compression in order to get something that works seamlessly. When you get it right, it can just add a body to something where compression or limiting would have just been like a blunt instrument.

Q: Can you maybe give us an example or two of what you mean by unusual pumping effects? Is there something that jumps out as a specific thing in that respect?

It's like most compression: it's all down to the attack and release time. Essentially, what you're doing if you're doing it "in the box," for example, you copy the audio down onto another stream, you put that into a compressor or a limiter, and you hit that limiter really hard, then you're mixing that result back into the main unadulterated track. Because that compressor/limiter is being hit so hard, sometimes you'll find that the release of the compressor will end up having more of an effect than you would have normally expected, and that's where you'll have to ride the output of the compressor/limiter so that you're also manually riding the peaks and troughs. Then, when it does suddenly release and cause pumping, you have pulled down the output level so that the effect isn't noticed. In other words, it gets lost in the main body of the track again.

Q: In your opinion, how does remastering differ from what you might call frontline mastering? What are the different skillsets you need when you're working in the remastering realm?

It depends on your reason for doing the remastering. I was saying earlier about compilation albums, regardless of the time period, you're trying to put together 20 mastered tracks done in different places at different times, so they will all sound different to one another and wouldn't necessarily sit well together. In that case, you have to do an element of remastering in order to achieve that consistency across the number of tracks. The thing is to do the minimum amount needed in order to achieve that result. It's not in the brief to go, "Oh, well, I can master that better than he did, and I'm going to do all of this." The brief is that the producer, the label, the artist, whoever, were perfectly happy with the track the way it sounded. It's just that in this context it doesn't sit well, so you just do the minimum that needs to be done so that the whole release works well.

Other types of remastering can be an older recording that is going to be re-released. Sometimes the brief is that it needs to be as faithful to the original release as possible, and sometimes people want it to be made more contemporary. Over time, the contemporary sound changes; for example, disco was really bright, and then it drifted into really bassy. Then guitar bands came back in, and it was all midrange guitars clanging out. The whole thing does evolve; you can take anything from any given period, and what was a contemporary sound may not be the same as it is now. So you may be asked to make it sound more contemporary—that might involve more bass, or taking down the top end a little bit, or whatever. What I don't think you necessarily need to do is to squash the living daylights out of it.

Q: Yourself and Matt are very busy cutting these days. From your experience, what's the balance between cutting work and standard digital work for CDs and online releases?

I think it's probably about 50/50, but generally with anything that is going on vinyl, there will be digital associated with it. What I have been working on are jobs where the client wants their tracks cut half-speed, and so obviously I'm only involved in the vinyl aspect of those tracks.

Q: Because this is quite a contentious issue in the sense that you might do a particular master for CD, a particular master for mp3, are you doing multiple formats and different masters for everything? Or are you not doing that?

I would view it as, there is a digital master, and there is a vinyl master. It's quite annoying that a lot of people just assume that the vinyl is just done from the CD. That's certainly not something we would endorse here. It's quite depressing when you have some clients that are opting to have the cut done in the factory from the CD, because the vinyl release could be so much more.

Q: What are the risks for the near and distant future for you guys and everyone else who's using those lathes?

Well, Sean Davies is cutting back his workload a little. There's Crispin Murray and John Goldstraw and the guys at Audio Related Technology mainly doing the servicing, so they're certainly not allowed on an airplane together! If you were looking at it rationally, the whole thing is so precarious. We have three brands of lacquer; two of those three are made in the same factory, so that doesn't really count. We have two manufactures of cutting styli. The factories are absolutely heaving with work and having long lead times. The bottleneck in the whole process seems to be the galvanic process, the electroforming process, and although GZ are investing in new presses they've designed and built themselves, I don't know of anybody that's making any serious inroads into creating more metal working facilities. If they did, I don't know that the factories would actually be willing to take the metal work, but I think that that would be a good avenue to pursue. So as I say, the whole thing is pretty precarious, and I think it will stay that way until more people are convinced that the vinyl resurgence isn't a one-off, but that it's here for a while.

Q: In terms of the broadcast wav and ISRC numbers, could you explain how that all came about, and how you got involved in it?

I was on some compilation albums and would receive the audio from the label. I would compile the album and send it back. Then you find out that they've sent you a wrong version. It can be wrong, for example, because it's the explicit version and it should have been the clean version, or the other way around. You can see from the label copy that you have the right title, the right artist, and you know it's the right duration, but you don't know for certain that it's actually the correct mix.

It just occurred to me that if associated with the audio was an ISRC, then it's not beyond the wits of a programmer to make the ISRC display within the waveform of my workstation, such that I can look at the label copy ISRC, and I can look at the waveform ISRC, and if they don't match up, well, then I now know I've got a problem before I've wasted time.

I understood enough about broadcast wavs to know that the broadcast wav facilitated a lot of extra information that wasn't possible to put into a conventional wav file, and I knew that the European Broadcasting Union [EBU] created the specifications for broadcast wavs. I found their website and asked who was managing the standards for bwav. I got put in touch with a guy and explained what I was trying to achieve in terms of a unique identifier, and he got it straightaway. They just picked it up, ran with it, and came back with a spec. After that, it was just a case of constantly pestering all the software developers and getting them to include it. I went for the big three mastering DAWs: SADiE, Sequoia, and Pyramix. I set up an email group to ensure that they were all talking to one another, because I had made it clear that it had to work cross-platform. They collaborated and issued updated software, and then SoundBlade came on board. When the specification was published, other people who had seen it realized that they had to be doing it as well; it has snowballed from there.

Q: Can you explain again about how labels are throwing it up, and are they actually recreating those bwavs with that information? Or is that something we as mastering engineers are always going to have to do? What's your vision for it in the future?

There is free software available to take a conventional wav, to add an ISRC, and generate a bwav with that ISRC embedded in it. It's not as if the mastering community is trying to create a standard that then guarantees them extra work. It's something that is a good idea, and because it's a good idea, it has to be seen that nobody is trying to an advantage with it. I pitched it to the Boards of AIM and the BPI; they understood and said they'd recommend it to their membership. At the moment with the major labels, they are yet to adopt it. However, independent labels are all asking for it. I believe that, once you identify a track because it has the ISRC number in it, you can start exploiting that information. Broadcasters can start automatically doing accurate reporting of playouts, because it makes it easier to report sync usage. The master with the ISRC, when it's sent to a digital aggregator, they can adapt their encoding software to pull the ISRC information into the encoded file; hence, it cuts down the amount of re-keying of information and the potential for errors to creep in. And ultimately, although I don't think that there will be a single ISRC database, it provides a keystone for a database to be created around that track. The ISRC is a dumb identifier, much like a barcode, but when linked to databases can help with royalty payments, writer, producer, performer, and engineer credits. You will always know that you're talking about that track.

Q: Do you think the "loudness wars" is over?

Not yet. When I was a mastering engineer at Abbey Road, I did the mastering for Oasis—*Definitely Maybe* and the singles, "Some Might Say," "Roll With It," and "Whatever." They were the band that were going to be louder than anyone else. They were [signed] through Creation and Sony, and Ray Staff was working for Sony in London. He phoned me up, because a guy from Sony in America had complained about how much compression was on the masters, but the whole thing about it was that it was the USP [unique selling point] of Oasis: it was going to be squashed and loud. I look on that as being no different than using a vocoder, or echo, or anything else. That was what they chose to go with, and not everybody had to then go and do it as well. For me, it's the equivalent of using a vocoder on the track, and then everybody else putting a vocoder on their track! However, it did kind of feed into this paranoia of not wanting your music to sound quieter than another track in a club. R-128 is now adopted throughout the whole of UK and most of Europe. Soon people will come to realize that a heavily squashed track that is then level corrected in R-128 sounds more diminished than a track that meets R-128 without having to be reduced in level at all. Once that starts happening, and they twig that, then we'll stop being asked to do it. I find it really annoying that mastering engineers get the blame for squashing a track, and it's not something that we choose to do. We do it because we're asked to do it. So, it's something that is extremely frustrating, and it's something that's parked firmly with us as being "to blame," but as I say, it's not an automatic: we do it because we're asked.

Q: Are you being asked less often now?

Not at the moment, but Spotify, Apple, Tidal, and YouTube are all using similar methods of volume adjustment, so it's starting to spread outside of traditional national broadcasters. People will start to notice that, "My track quieter, why? I've squashed it as much as I could?" Eventually they'll figure it out.

Q: How has mastering changed since you've started?

It doesn't come in on reel-to-reel anymore.

Q: Does that translate into different practices, different client expectations?

Yes, there's a lot less checking going on than there used to be. The next thing that I'm trying to push through with broadcast wavs is for the inclusion of a checksum. At the moment, we throw files around in the production process willy-nilly, using umpteen different providers, and if it plays, we assume everything is okay. It's not beyond the wit of man to use a checksum, and it's something that should be in as a matter of course. So that as you import a track, it will tell you that something has failed the checksum; you can still listen to the file, because circumstances may determine that you have no alternative but to use what you have been sent. However, it's also flagging up and telling you to check this carefully, and if you can see or hear that there is a problem, then you can go back and ask for it to be resent. At the moment, if it plays, it goes, and that's no way to be working.

Q: Where do you see mastering heading, in relation to what's going on in the record industry at large?

I think that mastering is becoming more of a "cottage industry," and by that I mean that there are a few mastering engineers now who have built their rooms in their back garden. I just think the economy as a whole makes it really difficult for big facilities to exist. I ran a big facility for a while—it's nothing but a headache. So the people who are running Metropolis, the people who are running Air, hats off to them. If they can still keep going, they're doing a bloody good job. Abbey Road is a different kettle of fish because it's part of Universal now, and I don't think that's necessarily a bad thing. Universal can get their money's worth out of Abbey Road, but for a big independent facility, I think it's tough going.

Q: Do you think maybe that's a potential future trend, to move mastering back "in-house," as it were?

Possibly. You tend to see within the music industry, things are going through cycles. The whole of the music industry was in Soho, then it moved out west, and now apparently Universal is going to be moving east to King's Cross. Sony decided studios didn't make any money, so they closed their London studio. Decca closed their studios. It's not beyond the bounds that they would decide to move it in-house; after all, Sony have opened a vinyl factory in Japan. However, the only way they can do that now is through some sort of acquisition, because as we covered earlier, there are a finite number of lathes. And if you're a major label, and you want to have an in-house facility, it would be ridiculous to entertain doing that without the ability to service vinyl, which means you've got to talk to somebody who's got a lathe.

Q: What do you think, if anything, the "resurgence in vinyl" points to with regards to listeners? Why are people buying vinyl again?

I think there are a vast number of reasons. There are quite simply the people that want to buy the picture disc or the sleeve as a piece of art, there are the people who like the way that vinyl is more tactile, and there are the people who buy it because that's the cool thing to do, and they put it on their turntable with its USB output. But then there are also the people who recognize that there are engineers like us who will do a version for digital and will do a version optimized for vinyl, and a lot of people like something that is a little less compressed, has a more open top end to it, and they just enjoy that openness. I hope that there will be more and more of them!

Q: If you could go back in time and catch yourself just as you were about to embark on your career, and tell yourself one thing, what would it be?

This is the best time ever, and don't forget a minute of it!

Q: In regards to your room, tell me about your speakers: have you got KEF's?

Yes, they just sounded right in the room.

Q: Did you try others?

Yes, a few. What happened was this: when I first started cutting records, I thought I'd better have a record player. So we went to Laskys, and I spent the entire budget that my wife had set aside for a system on a pair of KEF speakers. They've stuck with me through all the different incarnations of Alchemy and been used in one way or another. I had them in the room when we were building it. Matt and I thought, you know, these sound really good, but they just don't have the bass. So I got in touch with KEF, and they wheeled in a bigger pair, and they had everything I wanted, with the extended bass that the other ones couldn't achieve. I don't understand rooms where they have a particular speaker in every room, like Abbey Road have B&W throughout. Equally, at Metropolis they've got PMCs throughout, whereas I'm actually a believer that once the room is there, you find the speakers that work best in the room. Having a preconceived thought that, "I'm going to use these" is probably a bit narrow, and you have to be open to try different things in order to see what works best.

Mandy Parnell

Mandy Parnell is founder and senior mastering engineer at Black Saloon Studios in London, England. She has worked on projects for a wide range of artists including The XX, Feist, Sigur Ros, Björk, Frightened Rabbit, and Brian Eno, to name a few.

Q: How did you become a mastering engineer?

It all started with a friend of mine who was a housekeeper at the Manor Recording Studios in Oxford. The Manor was a Virgin Record Company owned residential recording studio. This was back in the '80s. The bands she looked after included Rush, Queen, and so on. It was a big beautiful mansion house and studio. I would be invited to go down and visit on the weekends to hang out with her. I was 16 then, and I had a strange life.

I was a runaway and ended up in a children's home and dropped out of school. I was just doing odd jobs; I didn't really have a career angle at that point in my life.

I was down there at the studio one weekend visiting my friend and sitting around in the kitchen when the assistant engineer said to me, "Do you want to see the studio?" I was like, "Sure, it would be great to have a look." I walked into the studio, and all I can say is that I had a "clicky" moment: something just clicked inside me, and I started asking him a lot of questions about how it worked. I'd been listening to records since the age of 5—I truly was obsessed with music. I never really wanted to play music, and even though I did learn to play different instruments, I was just obsessed by finished records.

I came back to London and looked for a course. Back then there were three courses in the UK that I could find. There was Surrey University's Tonmeister course and two other private colleges that were doing music tech and production courses. The School of Audio Engineering (SAE) had just come to London. I think they were in their second year at this time, and we had another school in South London called Gateway. I joined up with SAE, because there was no way Surrey would accept me with no qualifications, and Gateway was too far to travel. I had to work very hard through SAE's course. After completing the course, I went into assisting in recording studios for no money. I was living with a singer, a friend of mine, who used to buy *Music Week* magazine. I never read *Music Week*; I still find it quite a hard read. I was on the phone to a girlfriend, and I started flicking from the back page through and saw an advert for assistant mastering engineer at The Exchange. I called them up and was told, "We're not taking any more CVs from tomorrow, so you need to get your CV up here today." I didn't have a CV typed out; I could not access a typewriter, and this was pre-computer days. I handwrote one with spelling mistakes that I scribbled out. They invited me in for an interview as more of a joke to see who would have the audacity to send such a CV—they actually interviewed me because they were intrigued. During the interview, I was informed that they wanted to also hire a female with the guys they were hiring then; I got the job. That was the start of it.

Q: How did you learn to master?

It's really weird, because I have a lot of interns and assistants that come and work with me now; I also do mentoring for quite a few young people. What I say to my assistants is, "Nobody ever really taught me how to master, and I will not be able to teach you." It's a long process to learn mastering. We started off with two weeks in the copy room and a week sitting in with an engineer. We'd shadow the head engineers, and we couldn't have asked for a better start shadowing Ray Staff, John Dent, and Graeme Durham. Nobody really taught us how to listen or how to EQ. Literally sitting in and getting your ears accustomed to it over time was how we learned. In the copy room we were doing tape transfers, so it would be reel-to-reel tape copies, and we'd do the bin masters for cassettes. Then coming onto the CD, when the CD started to really take off, we would do all the production parts and all the cloning for CDs. It was a slow process and all in real time. When we'd sit in, we would be taught about cutting and how to use the machines. We'd be taught what you could and couldn't do when cutting vinyl. Again, you weren't really shown about EQ from an artistic approach, only technically. They took a very hands-off approach artistically: "You will figure it out by yourself." I suppose they realized whether you were "in tune" or not by the questions you asked. I think I shocked them when I would take the manuals home and read them. I was a geek girl, and I wanted to understand. If I didn't understand, I wasn't confident. I needed to understand what was going on and the equipment that was surrounding me.

I was at The Exchange for quite a few years before I left and went to America, got married, and had my son. The reason for that was because The Exchange expanded very quickly and took on more assistant engineers. They grew too big too quickly; they then realized that they'd grown too big and had to make people redundant. It pushed me back into the copy room, and it meant I couldn't move up to become a junior mastering engineer. I wasn't really willing to sit around waiting for the day to move up again. That's when I went off to New Orleans to live. When we moved back to London, about four or five years later, I got my job back at The Exchange and walked back into the studio as a mastering engineer.

I didn't get a formal training, in a sense. I don't think it's about that, and that's what I say to my assistants: "For the first year you're not really going to touch a button on my desk; I just want you sitting in the room. You're part of the production path, running production parts and learning critical listening; I just want you sitting in. If you've got questions, ask me and I'll explain stuff to you as we go." I'd just talk to them about how they should be listening, because for me, I think when you're mastering tracks, you listen in a very different way to people who are mixing.

Q: What are the things that you typically listen for, and what aesthetic areas do you address regardless of genre, if at all?

I work the same way whether clients attend or not. It's easier for me when people attend, as I can ask them lots of questions about the project. At the start of the session, I'll get them to talk about the project and what direction they have taken. If it's an album, I'd prefer it if they had a running order. I'd play the whole album with the running order, listening and not paying much attention technically. I won't necessarily be sitting in the sweet spot listening, either—I'm listening to what's going on emotionally, the flow and continuity, and I'd make notes along the way. I normally sit with a note pad, writing whatever comes to mind. I can write down emotional notes. It could be sonic notes. I could write down "bass" because I know I need to work on the bass. It could be "make it more funky" or "follow the vocal" or "follow the bass guitar" or whatever I feel, sadness, happiness, etc.

The first port of call is thinking about the box that it's going to fit into and whether the mix is good enough for that box. It's hard for me, as it's so different now to when I came into the industry. When I came into it as a mastering engineer, for me to reject a mix, it had to be pretty bad. The cost for the team of me rejecting the mix was huge. It meant they needed to go back into the recording studio, they needed to buy more tape—it was a lot of money for the team. For me, there are so many different styles of music, I have to think about the audience we're aiming the art for. For instance, hip-hop and R&B is quite a small sonic box that the art needs to fit and work into. So, if the mix isn't quite there, I would ask them, "Is it possible to go back and mix?" Today, with more bedroom productions, and a lot of mix engineers mixing "in the box," it's very easy to recall the track and give it a tweak.

It's still hard for me; I still have the mentality of only recalling it if it is really bad. It's a strange thing working with some production teams: this recall of mixes and rejecting them can feel endless. Often with younger mix engineers, I can hear that they have their licks, and if they have potential, I will try and help develop their style with their mixes. It's hard to gauge when to ask for a recall. If clients are in the studio, I will talk to them. With some people, I can just see by when I talk to them that they don't want to go back and address the mix again; it's too emotional for them to open it up. So then I just do what I can do and leave it open for them. You need to think about the art first, then the box. That's your starting point of where we are going. If it's a new band and they're looking to get signed with a record label, I try and look at the project with that in mind. I approach it in that sense of playing it on some really crappy speaker, especially in mono. If they want it specifically for the dance floor, if it's got to be a 12 inch record, then you might view it slightly differently and approach it with that in mind. You think about where it's going, Who's the audience? What's the purpose? How will it sit against other music in different genres?

Q: What are the tools you use to fit things in their respective boxes? Do you have a preferred chain, and can you take us through it?

Again, it really depends on the session. My attitude is that every piece of art is different. When I'm doing the listening, I'm listening to the mixes, and then I'm thinking about reference points. The reference point might be something from my past, something I've heard, something more current. If I'm not sure, I might ask the client what their reference points have been. Often, I might have to get into what has been their inspiration for the album? What artists inspired them sonically for the album? The mix engineer and producer might also have reference points. I try and keep an open dialogue about what they want. Then with my equipment toolset, I have different software platforms that I play the files from. Again, I might decide to use a different platform depending on what the project is, maybe what they've used. If they've done it in Pro Tools, I might use that; I might use SADiE, I might use Sequoia, I might even use Logic—it really depends on what the project is, what they've done, and what I hear.

My normal set up is to start off with the SADiE Mastering software going into a Prism ADA8; out of there, I may go into a limiter/compressor, into an EQ, into another EQ, then it comes into my desk; I have an EMI TG Transfer Console. I may use plugins at the beginning of the chain, more for doctoring the mix. After all my analog chain, I convert it back into digital with the Prism ADA8. Then I have a TC 6000 and an L2 limiter, so I might pick one of them, I might use both of them, I might use a plugin limiter—it really depends, because each job is so different. I have a very rare Inward Connections EQ that I use as well; I've got loads of tools. I've got a Hammer EQ that sometimes comes out. I have my EAR 660 compressor/ limiters, I have the Prism Maselec and a Phoenix. So it really depends, again, on what I hear, what I'm going to plug in, if that makes sense.

Q: How about monitoring?

ATC SCM100A SLs with the sub. I've got a pair of old 15 inch dual concentric Tannoys with Barefoots, and I've got Meyer Sound HD-1s and a pink pair of Reftones.

Q: How is your thinking different in regards to the different formats, if at all?

It depends what it's for, whether it's just a CD release or digital release. Again, I'll speak to people about what they want. It's really hard at the moment, as I think we're in a really strange place in mastering. Everyone says that "we're becoming more dynamic and we shouldn't worry about the loudness wars," and all this stuff. But I've discovered over the past, I'd say, year and a half to two years, everyone wants it louder than they've ever wanted it in my whole career. It's really odd considering we're trying to go quieter, especially with the loudness normalization being implemented with the different digital platforms. Unfortunately, lots of people are still doing very loud stuff through these platforms. I will refer back to the production team about what they're looking for from their mastering and how loud they'd like their project; there are no standards anymore. If you're mastering for vinyl, you generally master a more dynamic version. I'll master in high-res if I can, and then we'll cut the vinyl in high-res. It depends on the platform it's going to and what the team wants as to how I'll approach it, but it would be different masters.

Q: Have you had any great learning experiences for you along the way that stick out as, "On this project, I really learned 'this' about mastering?"

Yeah, Björk. It started off when she had an album mastered by another mastering engineer, the *Biophilia* album. She wasn't happy with the project. The record company who I've worked with for my whole career, One Little Indian, approached me: "Can you cut the vinyl? Because Björk didn't like what has been cut, she thinks it's very sibilant." I said, "Sure, send me the files, and I'll do it for you." Then they came back and said, "Actually, could you redo the CD as well, because she's not happy with that." So I was like, "Okay, hmm, something's odd." Then they said, "Actually, can we fly you up to Iceland for playback with Björk, because she's in rehearsal at the moment putting her live show together, so she can't travel to London." So I was like, "Fine, I'll fly out there." Then they asked, "Can you take a mobile rig?" I was like, "Okay, that's a really tall order." I have done it with some of her friends, and this is where she'd heard about it. It's a big ask—we're talking about Björk here, who I have not worked with before. This is a very big release for her. It's the one that she did the iPad apps for, it was a big project. So I said, "Look, all we can do is try; let's try."

I flew up to Iceland and went to Addi 800's studios, we moved all the speakers around and rearranged the room to set up the equipment. If anyone came into my room and did what I did to Addi's room, I would have probably told them to leave. He was very humble and let me do whatever I wanted. It was so sweet; I found all the guys up there were great. They'd all got together and called in all the mastering equipment that was available in the country and brought it to the studio, so that I had a pick of all this equipment. It was incredible; I was like a kid in a candy store. We set up the room, and Björk's assistant came down before Björk came in. He brought the drive and opened up a Pro Tools session with the mix. So I'm looking at it and watching, and I'm like, "Well, that's the mix there." And he's like, "Oh . . ."

I asked, "Well, haven't you got the stereo print?" And he answered, "No?" And I said, "Well, can you bounce it? Can I work from the stereo?" He said, "No?" And I asked, "Why not?" And he said, "Because it's not finished." And I went, "Well, who's finishing it then?" And he said, "Well, you are, with Björk."

I went through it in my head; it was about 15 minutes before Björk was arriving at the studio to meet me. I sat there and I thought, "Oh my God, what do I do? Do I just say I can't do it? What do I do?"

Taking some time to think, I considered that, "Okay, we'll just have to go for it." We were on a really short deadline. The record company said to me that we had to deliver it to Japan on the Friday, because she had already pushed her release dates back so many weeks. We had to make it work. We had the craziest four-day session with not a lot of sleep. Björk and I worked very closely finishing her mixes off and mastering the record. She actually taught me not to be scared of things, to follow your instincts and to just follow it, and to listen, and you could do whatever you wanted, so that's what we did. She really opened my world up to actually not being scared about the mix, if that makes sense. Not being scared to speak my mind a lot more openly. For me it was a big turning point in my career, doing that project with her. She was very cool. We did her last album the same way, actually.

Q: So you finished off the mixes and mastered them?

Yeah, we got all the stems and finished everything. I changed lots of small details and did lots of editing, recalled vocals, and recompiled vocals. We did lots of crazy stuff.

Björk is very humble with her team doing stuff. She's working with mix engineers, and producers, and beat makers, and she will let them do their thing. Then she will put it back to where she wants it after they have finished rather than going in heavy and controlling.

She's incredible with her team, and we all worked really hard together on this last album. The first one we worked on was just Björk and myself, whereas this one was a bigger team. I flew up on Boxing Day, and we worked through Christmas and New Year's. Of course the rest of the team were with their families, so we were talking a lot over the phone, and they were sending bits to us. There was a lot of communication even though they weren't there.

Q: Do you work with stem mastering at all?

Yes, but I try to avoid it. I do try to get the team to sign off if they can. On some art, it's really hard. I find with experimental music it's sometimes harder to sign off on the stems because you're not sure what's going to happen through the mastering. If it is attended, it is easier, as the artist will guide me. For hip-hop and R&B, it's easier if I have the stems. Often people don't mix the vocals loud enough, so we need to get in there and change that; it depends on the genre, really.

Q: So you'd rather them have it nailed in stereo, if you can?

I mean, I've got one at the moment where we've mastered everything and everyone's really happy. We're very late in the day, and we're just about to run production parts, so everything is signed off, and then the artist came back and said, "I'd like a vocal-up version." But all I have is the instrumental and the vocal, and I have the stereo mix. Anyone knows that if you put the instrumental and the vocal together it doesn't equate to a stereo mix. So he wants me to do a vocal-up version, and I said to him, "That's not a problem; it's just not going to work the same. So I'm going to have to go in and master it from the stems, which will be different, which will cost you more money." I said, "Can't you ask your mix engineer to pull up the mix and just bring the vocal up? It would be way cheaper than getting me to do it on the stems." It really depends on the project. I'd rather not get into stems; it's a lot of time for me. It makes me have to think of the mix first—I have to put a mix engineer hat on, so to say, and then come back to it with a mastering mindset. The two things are very different, I think; well, they are for me, anyway.

Q: We've heard that, because of decreased budgets, timelines for mastering projects are receding. Is that your experience?

Yeah, everyone "wants it yesterday" now. We're in such a fast-paced world with the internet and the delivery of files; everything is so quick. When I came into it, all the tracks were on reel-to-reel tapes. You produced your master disc and it went off to the factory. It was phone calls rather than emails; everything was a lot calmer. It was also a longer time frame to the release of a record. Now everything's shorter. It's really weird: digital is shorter and everyone wants it faster. Then we have these long timelines for vinyl from three to six months for pressing, which is really confusing, so you've got these two crazy worlds going on. Everyone wants it now, but when it comes to vinyl we have these very long lead times. We are given slots with the factory, and of course if you're cutting lacquer, we want to get it to the factory within 24 hours of cutting it, because we need to get in to processing within 72 hours. We need to get it in that time frame for the optimal sound. We're a very small studio and a very small team; we're juggling a lot to facilitate everything and everyone. Often, people book a session, and then they're not ready with the files and we need to reschedule, or we'll get an email or phone call and it will be a producer or artist's project and they want to come into mastering within two days, and we're booked up for like two months, but it is a session we would really like to do. It's really hard to juggle with when those situations arise—it's crazy. Everyone just wants it now.

On the budget side of it: yes, the budgets have got smaller, but they've got smaller on the whole project, from start to finish. As we know, the artists aren't getting signed for as much money. The recording budget is smaller as well.

Then we have this other situation where we've got so many schools now. When I came into the industry, which was 30 years ago, we had three schools in the UK. Now we've got over 200 churning out music tech[nology] and music production students, and there are no jobs for them. We're in such a declining industry. When I came into the industry it was the '80s, it was the heyday. There was a recording studio on almost every corner: and it was rock & roll. Now we've become a cottage industry. It is something I talk about when I go into the universities: "Where do you think your career is going to be if you start setting up a mastering room in the bedroom of your parent's house, mastering inside the box, which isn't very good? You're conning a lot of people sometimes with some of the stuff you're doing, and you're charging 15 pounds a track?" I couldn't open my studio door for 15 pounds a track. So I say to them, "Where do you think you're going to have a career if you keep doing that? You need to keep the industry professional." The record labels are a business. They're all about doing it cheap, you know, "How cheap can we get it?" When people are putting budgets together they say, "Well, we can go into that studio and get it for this money. Why do you want to go to that studio and pay that, when we can pay this?" The budget side is hitting us at lots of angles, you know. "Do a trial for free. Oh, we wanted to try you out for this album project, but we expect you to do it for nothing." I refuse absolutely point blank, because to "try it out," you actually—on the first track of any album or project—spend the longest time, because you're seeing what's going to work and how it's going to work. Why am I going to spend two, maybe three hours doing a try-out where I'm not going to get paid? Even if you spend all that time on it, you'd be lucky if they paid you 100 pounds for it.

If you're putting me in a competition, then I have to explore every angle. I do it with any job—any of my clients will tell you. I'm the same at the start of any job. I work really hard on the first track finding what's going to fit the project, and to get my reference point for the rest of the project, because that's what it is. The trial thing can be a bit hard, because they do expect it for nothing, and they get upset that you're not willing to do it for nothing. I have assistants and bills I have to pay. Part of it is that we do have students coming into that marketplace that will do anything to get their foot in the door, so they are giving it away for nothing. Like I said, the record companies are a business; managers are a business.

Q: What do you recommend that a kid who has this passion to master should do?

What I say to them is maybe for you to get the gig, have a contract: "If you're going to use the work that I do, then pay me a reasonable price for it." It could be after it is released that they get paid first. If they keep giving it away for nothing, they are not going to have a career.

Q: Does that also undermine, as well, in general?

Yeah. It's quite interesting when you present this situation to students, which I've been doing for the past few years now, to see them start to understand what is happening. It came about because I'd put an ad up for an intern, and someone got in touch with me and said that he was already mastering. I looked him up and he had a website, and he was charging 15 pounds a track. I just emailed him back and I said, "Well, I can't really let you become an intern with me because technically you're my competition, and especially when you're undercutting me like that."

I really thought about it, and I thought, "How are these kids going to have a career?" I'm going to be lucky, hopefully, and retire as a mastering engineer. But where is the future for these kids? If you look at them in their twenties and they're giving it away for nothing, where is it going to go? It's terrifying. What the colleges and universities really need to start teaching these kids is strategies to earn money, not do it for free. If we don't start doing that, there's not going to be a future. It's going to make our profession a hobby. It's not going to be a career, it's going to be a hobby—and is that what we really want for the music business?

Q: Is there less awareness of sound quality for artists when people are mastering for such small fees? Are they doing a good job?

Who knows? There are so many wrong views about mastering. People think that mastering first and foremost is just about making your tracks louder. That's not what mastering is.

Q: What is mastering? If you could educate people about what you do, what is it?

So what's the role of the mastering engineer in any project: it's a sounding board first and foremost for the production team, so that they haven't lost themselves in the project. It's a sounding board for the mixes: are the mixes working for the purpose that they intended them to? So that's the first meeting booked, we talk about where they're aiming the project for and the direction we're going to take it in mastering. The talk about loudness for me is probably the last thing that we talk about. We talk about what they're looking for sonically in their project, you know. We play back their whole album, and if the producer and the mix engineer are there, we'll talk about if it's translating well on our system. A lot of times now, people are mixing in bedrooms and in rooms that aren't designed to be a studio. There are often lots of issues with the bass, and they're not aware of it until they get into a mastering room, especially with inexperienced young mix engineers that haven't been taught how to mix. They don't know how to do bass management. They don't understand about phase and the problems with that, so stuff might not be mono-compatible, which is still an issue today. It's an issue that's come back around recently for me with the up rise of these one-speaker boxes like the Bose or the Sonos system which are pseudo-stereo: they have weird, psychoacoustic stuff going on with them. We're back like we were years ago with radio, how it needs to be mono-compatible. So there are lots of things that we need to look at in the beginning of the session before we even talk about the "loudness button."

I've had some major issues recently, with two very big projects that I can't name: top mix engineers and producers worked on these albums, the record companies are playing it back on one of these one-speaker systems, and I've received phone calls saying it doesn't sound right, yet everyone inside the production team has signed off on it because they've been playing it on stereo-compatible equipment. When you start investigating, the issue is that it is not mono-compatible. This issue made me understand why so many record companies were coming back and asking for projects to be louder, because it's not working through these boxes. So many of the record companies have got them installed in their offices, it's ridiculous. For years, most people would play it on their laptop speakers, so they were stereo; there was quite a bit of space between them. When you look into the one-speaker boxes, they're really not stereo. I've had to go back in and remaster tracks and basically master in mono to make it work through these systems, causing the stereo version to sound too hyped.

Q: You mentioned you've got an EMI Mastering desk. How important is that to your process?

It's not; it is just a tool. It was an interesting thing, because it became apparent through years of clients saying to me, "I can actually hear your sound. When I hear a record, I sit there and guess who's mastered it, and I'll guess that it's you, and when I look, it is." It was just an interesting thing. So, I started going back and listening to stuff and thinking about the sound. It's the same for all great mix engineers and great mastering engineers when you talk to them—it's not about the tools, it's about the ears, and it's about how we interact with sound. We will try our best with any piece of equipment because we're looking for a sound that we perceive we are hearing in our head. I remember about six or seven years ago, I had an artist come in and went to play the files—he had flown in from Germany. We went to play the files, and something went wrong on the desk, and I couldn't get a sound out of it, and I couldn't fix it to continue with the session. So we did the playback of the album, and I showed him what I was thinking of doing inside Logic, just with plugins. I just said, "This is what I'm thinking about, this sort of colour, different textures, whatever." My assistant at the time said, "I can't believe that you managed to do that with Logic." I just thought, and I looked at him and said, "It's not about the tools, but the ear set." I can show people what I'm looking for, and I can give a percentage better, I don't know how much more, 10%, 15% better, maybe 20%, I don't know. But I can give a view of what I want to do through anything. I think you can master with anything—well, I have done. It's about the ear set. You do your best with whatever you have. I'm sure all the people I've looked up to through my career say the same. I mean my desk is beautiful, I love it. It makes my job very easy. It's an easy tool to work with, but it's a tool. It makes it easy because it's very quick to get what I'm looking for out of it. With other equipment, I'd probably have to work a lot harder to get what I want.

Q: How did you decide to set up your own room: Black Saloon Studios?

I was approached by Universal Mastering in LA to go over and to take over their Studio A. They had just built their new mastering rooms there. We had meetings over the course of about eight or nine months, and we were setting up the dates that I was going to move to LA. At the last minute, I just changed my mind. I don't know why, I can't explain why. I didn't feel comfortable about it; I wasn't sure it was the best thing for me to do. No other reasoner than I didn't feel right about it, so I didn't do it. That's basically how the studio came about.

By that point, I'd learnt the room, I had worked on some big albums that had done very well out of that room. I'd been talking to Sean Davis, who has been one of my mentors throughout my career, about doing acoustics and looking for rooms with me. I had a meeting with him, and he advised me, "Look, you're doing great work out of here, so why change anything?" That's basically how it came about.

I wouldn't recommend to a lot of people to set up their own businesses—it's hard work, it's really hard. As an engineer, you're facilitating artists; you need to get into a very art brain. Then you have to come out and do the business stuff; it's a lot of jumping around of different brain activity. Well, this is how I feel: I like to stay inside the artistic brain, not the logical brain.

Q: Are you being a little bit facetious, or are you serious?

No. I'm being serious. I think it can really upset the balance of what you're doing art-wise, especially when you're a small business, if you think that I had no business plan, it just happened very quickly. You're trying to build a studio with staff, and running it, and work. My career since I've done it has just gone absolutely crazy. Winning awards and Grammys and God knows what. It's just crazy, and incredible, but it's just a very hard balance. I'm all about being in the studio and doing the art stuff. I don't really want to think about other things.

Q: Are you cutting more vinyl, with the vinyl resurgence?

I'd say it's the same for me, but then I did come from The Exchange, where I was cutting a lot of 12 inches for the dance floor.

Q: If you could go back in time and talk to yourself right when you were first entering this career, what would you tell yourself?

Travel the world and then go and study law.

Darcy Proper

Wisseloord's resident American, Darcy graduated with honours from NYU's Music Technology programme. She began her career at Sony Music Studios in New York City in the classical department, later focusing her skills on mastering and broadening her musical scope to include all genres, from historical reissues to cutting-edge surround releases. After 20 years in the big city, she made the jump to Europe, accepting a position as senior mastering engineer at Galaxy Studios in Belgium. In 2010, Darcy became part of the team of adventurers dedicated to the revival of Wisseloord, where she currently serves as their director of mastering.

Over the years, Darcy has been honoured with three Grammy awards and nine nominations and has won several other awards for her work. She has had the pleasure of mastering historical reissue projects for such prestigious artists as Billie Holiday, Louis Armstrong, Frank Sinatra, Dave Brubeck, and Johnny Cash. She has also worked on stereo and 5.1 front-line releases for many talented artists including The Eagles, Ilse DeLange, Donald Fagen, Anouk, Porcupine Tree, Clouseau, Peter Maffay, Patricia Barber, Alain Clark, Jane Ira Bloom, Marcus Miller, Jef Neve, and Ozark Henry.

Q: How did you get started in mastering?

Like many people who end up in mastering, if you'd asked me when I graduated from school if I was going to be a mastering engineer, I would have said, "No, of course not," but life is full of surprises. I went to NYU, where they have a bachelor of music programme in music technology, and was one of the graduates in the first few years of the programme, I think. One of the benefits of going to that school, despite the fact that the programme was quite new, was that it was in New York City.

While I was in school, I did a number of internships—in an electronic music studio, a music publishing house, in live sound and theatre sound, and that kind of stuff. I really thought that when I came out of school I was going to get into the very technical side of things, maybe eventually research and development. My first full-time job was as an assistant studio technician in a dance remix studio. While that was still my full-time gig, a friend of mine was working for Sony Classical, and at some point they were looking for part-time quality control engineers with flexible hours. Extra income was always welcome, so I jumped at the chance.

At that time, 1630 digital tape masters had to be delivered for all the versions of an album. Our job in QC was therefore to make six copies of all of these tapes for the various pressing plants and then listen through them to make sure that there were no technical or audio errors on the tapes. We were just a list of people who on a given evening might be called up and asked if we could do a four-hour shift or so, listening. That was my real introduction to critical listening. I was eventually offered a position at Sony, leaving the dance remix studio. I was still doing a lot of technical work at Sony as a technician on remote recordings and gradually was also somehow sucked into actually working on the music. I hadn't really thought after school that I would be the one sitting in the chair working on music. But through doing this quality control work, which is sort of the lowest job in the whole facility, I learned how to listen. I then started working on digital editing systems—Sonic Solutions was the platform at the time—which was still an innovative idea. A lot of editing wasn't being done on workstations at that point.

By virtue of learning that, I suddenly became useful for client sessions and that kind of thing. It was first doing classical editing for frontline releases and doing classical reissues working with one of the older producers. He didn't really have any interest in knowing the computer end of things, and that's about all I knew. Between the two of us we made some nice reissues, and I learned a lot. That kind of work expanded through the years to include reissues for other genres for other labels at Sony. At some point, surround sound began to develop, and because I was used to doing multichannel editing and that kind of thing for classical, I was one of the few people who expressed an interest in working in surround at that facility. Eventually, between the combination of doing the reissue work, the quality control work, and being willing to jump into surround, I found myself attracting a broader base of mastering clients. In a way it happened on its own, just by taking the various paths that were presented to me and following my skillset and natural inclinations.

Q: How, then, did you learn some of the more aesthetic aspects of mastering?

A lot of it came from all those hours of concentrated listening and doing literally nothing else while doing quality control—focused listening to frontline releases and reissues of things from way before my time in a wide variety of genres. Also, eventually working on reissues and having to use modern technology to match the sonic effect of what had come before. You know, listening to the original release and trying to improve on it, but "improving" in a way that didn't change the character of the listening experience. It was a great learning opportunity working with these reissue producers, many of whom had done the original recordings in the first place. They could help me hear things the way they were listening to them. I was getting the benefit of their guidance and expertise as we worked together. I operated the new technology, but they let me know what we were aiming for. I very quickly developed a very broad sonic dictionary of what music was like over a period of decades.

Q: When you were doing the reissue work, did you find that you were working to preserve the production values from a different format and a different set of working circumstances for whatever the intended reissue would be?

I have to say, in the beginning, when no-noise or noise removal really came into the fore, we made a lot of mistakes and "threw a lot of babies out with the bath water," shall we say, by no-noising things a bit too far. Through recognizing the error of our ways, the goal was as you described—to give a better, cleaner listening experience for this material than was ever available for the end listener before, but trying to do it in such a way that didn't kill the original vibe. You don't want to destroy what was beautiful and wonderful about that era of music and the technology it was recorded with. I should also say that, of course, I was listening to modern music and was a fan of music in my own right, and not just listening to the old stuff.

Q: Did you find you worked differently if the material was new?

I guess the freedom when something is new is that you aren't comparing it to something else. You're trying to make it the most of what it is. With reissue, of course, you have to be very careful to respect the work that has come before, because so many people are familiar with that album as it was previously, and you don't want to disappoint the dedicated audience of that artist. With new material, you have the advantage of being able to just follow your heart. Things still have to be done technically correctly, but you're not so restricted by what happened previously. You can follow your own instinct and, of course, the input from the artists themselves about what they're trying to get out into the world.

Q: Is that a large part of mastering, is about being an expert listener in general? Knowing what you're hearing and having a vast body of knowledge, and knowing how to translate that?

I would say that's probably true. If I had to give somebody who wanted to become a mastering engineer just one piece of advice, it would be, "just listen to everything, and learn how to quantify what it is that you like or don't like. Quantify what you're hearing in what you're listening to." It's not to foster the idea that we mastering engineers know everything about listening. It also has to do with being able to communicate with artists and to be able to communicate with your clients. The broader your musical dictionary and the broader your sonic dictionary, the easier it will be when a client describes a feeling or something that they're trying to get across. I think that communication is probably one of the most important aspects of mastering. If you have a very limited sonic dictionary, then it will be difficult for you to work with clients that aren't limited to that relatively small sonic experience. I may have a client come in and who's working on some heavy metal music that's very modern in its approach, but they may make a reference to a jazz guitarist from 1950, some sort of flavour they're trying to hold onto in a particular section of a song or whatever. If I can relate to that, then that makes them feel like I understand what they're trying to accomplish and might be able to help them get there in the end.

Q: If you had to educate people about what mastering is, and why is it crucial to the process, what would you say? And what does it contribute to the production process—why is it essential?

I'm probably echoing Bob Katz in his book in that it's a two-part process. It's the last creative step in a music production and the first step in manufacturing.

As a mastering engineer, we tend to be able to offer perspective, because we haven't typically been deeply involved in the projects before they're ready for mastering. We tend to be free of all of the baggage that has been involved in the recording and mixing process. What I mean by that is that artists are so often involved in the technical parts of the process as well these days. By the time someone has written the songs, made the arrangements, and played and performed them, and then also been involved in the mix process, they're either at a point where they believe it's perfect, because they've done their very best to make it perfect throughout the process, or they're at the point where they think the whole thing is crap because they're just worn out from hearing it for so long. In the first case, where they think it's perfect, if they were to try to master it themselves or just put it out the way it is, then they might miss something that really does need to be adjusted or corrected or could be adjusted or corrected to make the music shine in its best light. On the other hand, when they are feeling very negative about it all, then the tendency is to fix things that aren't broken, which is, in essence, doing damage and making things worse. Being able to offer a fresh perspective and hear a project for what it really is is an important part of mastering. By making what are generally, in the grand scheme of things, quite small adjustments, mastering can give an album a feel of being a coherent, solid listening experience and bring the listener closer to the feeling in the music. The message of the artist makes sense because everything feels like it fits together in a way that expresses the artist's idea.

Then of course the technical side of the process is very simple, and that is the masters are put into formats that can be manufactured or distributed without technical flaws, dropouts and errors and that kind of thing. The artist can rely on the fact that the master that has left the mastering house is something that can be duplicated without putting thousands of copies of somebody's mistake out into the world.

Q: Can you talk to us a bit about what your process is? What are the things you do regardless of the genre of the material?

For me, I tend to be working in a combination of analog and digital. I don't think that's unique in mastering these days. It's a rare occasion where I don't use analog gear in the mastering chain. It could happen, but it generally doesn't happen. I can't think of a project I've done in the last three years that I haven't used some analog equipment in the process. As a technical description, I'm typically playing back these days from digital files on one workstation. I convert to an analog signal and run through my analog gear, which is the console, including EQ, compression, limiting, and a stereo widener, if it needs that kind of thing. Then it gets converted to digital at whatever my target sample rate is for delivery, recording into a different workstation, and I do my editing and final delivery from there. There might be some digital processing in the box when I'm playing back the initial set of files. Or there could be digital processing between the A/D converter and my final target workstation. It depends on what needs to be done. My goal with everything is to make it not necessarily sound like anything else but to make it the best of what it is. Whatever kind of music it is, whatever the artist's message is, I'm trying to make that message clearer, more compelling, and more emotionally accessible to the listener. If a track is a very intimate singer/songwriter kind of thing, then I want that connection with the story or emotion that the singer is sharing to be undeniable—that you cannot listen to this song without feeling whatever it is that they're feeling. In the case of a big heavy metal production, you should feel the energy and aggression. What is it that is special or exciting about this music that makes you inclined to listen to it? How I get there is generally by using this set of tools that I've described to you, making sure everything is technically correct with good metering and that kind of stuff. The rest is more about the emotional connection and figuring it out by doing. I may work to bring forward some element that makes me really connect to the heartbreak in the song or some other emotional aspect of the track.

Q: In the eventuality of a track needing mix alterations, how likely is it that you communicate with the mix engineer, as opposed to making adjustments within the master yourself and then just returning it?

There are a lot of factors involved in that decision—most of it is deadline and budget, probably. Is there even a remote possibility that if I did kick it back, that they would be able to do something about it? In the case where somebody has mixed something "in the box" and he or she could run it off really quickly, certainly I would ask that it be done. In any and all cases, if there's the possibility for it to be fixed in the mix stage rather than be attempting it in mastering, then I would prefer to do it there. Mixing and mastering are obviously not the same process. For the same reason, I don't believe that mastering from stems gets you the same result as mastering from a well-done, full mix. I think particularly in the case where the mix engineer has chosen his or her bus compression well and really worked to create a good feel in the mix, a lot of that is lost when you have to resort to mastering from stems. If there is a balance issue that can't simply be solved by someone having printed a vocal-up, vocal-down, and main version of a mix, then it's better addressed via more detailed mix adjustments. In general, the sonic results are better. That being said, when that isn't an option and there is something that I can do about it on the mastering side that will help bring things better in balance, then of course I'm willing to do that. I'm not going to say, "Well, here's the line in the sand between mixing and mastering, and I won't ever cross it." I'll do anything I can to make it sound better or, in any case, to make it sound more in line with what the artist would like it to be. So, if I can help, I will. But if the mix engineer can do something to fix a problem that did, in fact, generate from the mix stage then I would encourage that.

Q: Every mastering engineer seems to draw together their preferred set of tools that they work with in response to certain aesthetics within the music. Could you take us through and of yours?

The first thing that I try to do is to try and understand what the point of it is: what are we aiming for? That usually comes in a conversation beforehand with the client. "My goal is that it's the loudest thing on the radio," or "to have the best-sounding thing ever," or "to make my Mom real proud." Knowing what the client is aiming for makes a big difference in the direction you start heading in right from the start. If you know for somebody, for example, the audiophile audience is their primary audience, then you're going to handle something very differently than someone whose goal is to have lots of hits for their YouTube video and wants to be the loudest thing on the radio.

To understand my process, I think it's important to understand my approach to mastering. When I first started mastering, I had the feeling it was my job to correct things. That sort of puts you in a mindset where you're looking to find things wrong and then fix them. Over the years, and I think probably a bit out of self-preservation, I've changed my approach to listen initially for what I find interesting, unique, or compelling, or what makes this particular recording special. My working day is more pleasant because I'm focused on the positive rather than the negative, but also I'm not so tempted to dismiss things that are different and unusual, trying to fit that square peg into a round hole. What is it that I want to hear more of to accentuate that "specialness"? By virtue of maximizing what is special about something, you tend to minimize the flaws without dismissing its unique character as being a flaw because it doesn't fit in with the last project that you worked on or some other ideal. When I'm starting, I'm listening for what there is to love about the track that I can pull forward in the mastering stage.

Technically, I generally start off in Pro Tools, and I may be deciding at that point if I need to do any kind of digital filtering right from the beginning. For example, if there's way too much of the sub-lows and it's getting in the way, I may filter that off or do a bit of notching in a digital filter, right inside the box.

From there I have a very neutral sounding D/A converter—Prism—so most of the time I'm using that one to get to the analog stage. From there, to be honest, I don't have a huge arsenal of analog gear, because I do work in surround in analog, so I have to have three of everything at the moment, and we're doing a lot of 9.1 now these days, so I see that "problem" only getting worse if I want to stick to the "old school" approach.

In general I have two EQs that I call into play, and I may use one or both of them depending on what I'm looking for. I have a Dangerous Music Bax EQ [Baxendall Equalizer], which has a nice high pass and low pass filter, and then a simple tilt, let's call it a high shelf and low shelf option. That's usually the one I'll hit first, and I'll do the bigger sculpting there. Does it need some more high end in general? Do I need to do a little bit of a shelf from 230 Hz down because it's a bit muddy overall? From there I have an SPL PQ EQ, which is a parametric EQ. In that one I'll work on the details, assuming that the details need to be worked on after working in the Bax EQ. From there I make a decision between either my Millennia TCL2 Twin Com Opto-Compressors with the tube option, or I have a Neve 8061, which is a six-channel surround compressor, but of course you can use it for stereo as well. The Neve has a very typical Neve signature, so it tends to flatten things out a bit but also has that very cool Neve sound. When it works, it works, and when it doesn't, that kind of flavour is maybe completely undesirable. If you don't want pepper in your soup, grab the salt instead. With the Millennia, I find it a bit more versatile, and I can use that on a wider variety of materials. It can be quite transparent, but it does something. If it didn't do anything, then I wouldn't bother putting it in the chain, shall we say. From there I typically hit the Maselec limiter, one that also has a high frequency limiter. That can be handy, because you can put that in MS mode, where it's only working on the mono signal. If anything that I've done in the mastering process is maybe accentuating the sibilance a little bit, then I can use that high frequency limiter to put those "S's" back where they belong. That's also my safety limiter before going to the A/D converter. I also have a Dangerous Music Sum & Minus Box, which sometimes I use for a bit of widening. This is in the analog domain, where you can adjust the balance of the mono signal versus the side signal. If you put more side, it gets wider; if you put a bit more centre, it narrows it a little bit—so that's a handy tool.

The A/Ds that I use at the moment for most projects are the Lavry AD122s. From there, once I'm digital, I have an SPL Loudness Maximizer, which I think is one of the oldest boxes that does that kind of thing. I also have an AudioCube system, which has the higher resolution surround version of that same algorithm. I might kick that in if I feel it needs it, then from there I'm going into Merging Technology's Pyramix workstation. There may be some tools in Pyramix that I use for limiting, probably not in combination with but maybe in place of the Loudness Maximizer, if it's not giving me results that I'm looking for. Once I'm in Pyramix, that's my final destination before delivering for manufacture.

Q: Would you tell us a bit about your monitors?

I have EgglestonWorks Savoys. EgglestonWorks is a company out of Memphis. For the really geeky, it's not the new version of the Savoys with the new crossovers. We went to great pains to get the original crossovers, the more old-fashioned ones. The new ones I think are great for the audiophile market, but they weren't what I wanted for mastering. They're powered by Krell 400E monoblocks.

Q: Where do you see the mastering industry itself heading in general?

Maybe I'm foolishly optimistic, although spending so many years in New York hasn't really given me that impression; we tend to be a cynical bunch. As for the future of mastering, I don't know that I see mastering taking off and becoming the hot new thing or anything. I think mastering has always been about bringing the most out in the music for the format that is available to the consumer. We've gone through some dark years where what's been available to the consumer in a convenient format has been a rather awful-sounding format. In the beginning, actually, LP wasn't really a great-sounding format, given the equipment people had to playback the material on in their homes. Eventually that turned out to be quite a good-sounding format when the consumer technology was available. Cassette was a sonic disaster, but again, for the sake of convenience, it worked. CD became a great thing for decades. Then we hit this age of mp3s and highly data-compressed audio, which has painted a very bleak picture. When you can't hear details and depth, how could mastering matter, when mastering is largely about the details? It's a different story if you're just talking about adding 6 dB and calling it done, but the real art of mastering has to do with the details.

The good news is that high-resolution formats and large data files are becoming as easy to handle as MP3s once were, and consumer devices are available that make it as convenient to playback high resolution files as data-compressed ones. And physical media such as PureAudioBluRay support high resolution and surround/immersive formats as well.

As better-sounding formats become more available to the public, I think that mastering will hold its own. The point of mastering will always be to bring out the best in the music for the format that the consumer will eventually listen to. As the details are available in the end format for the listener, I think the importance of mastering will become more and more obvious.

Q: Do you cut vinyl?

I don't cut vinyl myself, but more and more of my clients are pressing vinyl of their releases. And more often than not, there is not a large enough budget to do two entirely independent masterings—one for digital release and one for vinyl. So I'm often able to convince the client to do a (slightly) more dynamic, less limited mastering overall to make it more "vinyl-friendly." This, of course, sounds better for digital release as well, so it's a win-win situation. There's still plenty of room for improvement on the loudness front, but every little bit helps.

Q: Finally, do you have anything to say about loudness normalization?

More and more streaming and digital delivery services are agreeing to use loudness normalization for their programme material. There are still obviously some discussions about a universal "standard," but in any case, it's a big step in the right direction on the technical side of things. Unfortunately, I can't say that it's made a huge difference yet in terms of clients wanting "loud" mastering. I've got all kinds of meters and can show them how the normalization will potentially affect their material, but those inclined to prefer "loud" over "dynamic" still generally choose "loud," very often for fear of not being as loud as "everyone else" on every conceivable release format.

I'm very much hoping this situation will improve as more people begin to understand the implications of loudness normalization and begin to take advantage of dynamics in music releases again. After so many years of "loud," there seems to be a whole generation of listeners who confuse "loudness" with "dynamic" and are somewhat numb to the difference. Many artists now use the two terms interchangeably when describing the effect they want their music to have on the listener. Thankfully, there are a few "pioneers" who are embracing the concept, and hopefully, as more artists become courageous enough to release their music with more concern for dynamics and musicality than for being stunningly loud, tastes will shift and others will follow suit.

Nick Watson

Nick Watson was surrounded by classical and jazz guitar music from an early age, thanks to his father Mike Watson's activities as a session musician and renowned guitar teacher. However, despite these positive influences, Nick chose the way of darkness and pursued a keen interest in rock and pop music, which encompassed the roots of rock and soul in the sixties right through to the glamorous '80s pop which dominated the charts in his school years.

In 1989, Nick graduated with a degree in electroacoustics and began his studio career, kicking off with a couple of years dabbling with production and recording before landing a job at Sound Recording Technology (SRT) as a mastering engineer. For the next 12 years Nick was kept very busy, gaining a great deal of experience working with an extremely diverse range of musical styles. From indie rock and pop to classical, jazz, and folk, artists, producers, and engineers would amaze Nick with their ability to turn out fantastic recordings on a very tight budget. . . . Nick describes his role at that time as running from quality control at one end of the spectrum to outright salvage at the other! In addition to working with new recordings, Nick also earned an enviable reputation for his remastering and occasional remixing of catalogue material, producing acclaimed reissues by artists such as Fleetwood Mac and Deep Purple and preparing repaired and enhanced masters for specialist reissue labels such as RPM, BGO, Snapper, Diamond, Sequel, Castle, and Connoisseur.

After a number of years, and via a promotion to chief engineer, Nick was then approached for a job at the Townhouse, where, as well as continuing his remastering activities, Nick's clientele broadened to include labels such as Sony BMG, Parlophone, and Island, working on recordings by Keane, Coldplay, Jamelia, Mika, and Gary Numan, among others. During this time, Nick also undertook mastering work in 5.1 and on the SACD format, and remastered seminal works by favourite bands The Kinks and The Small Faces, in the latter case garnering great praise from the remaining band members, who claimed their material had never sounded as good before.

With Fluid Mastering, Nick's excited about bringing together his experiences of major and smaller studio work. "A top mastering studio offers exceptional technical facilities," says Nick, "but too many great projects don't get the benefit, whether it's due to budget, or simply because clients find the major studios too impersonal to deal with. Our studio here at Fluid is the best-equipped room I've worked in, and we aim to provide every client with a service tailored to their needs and budget."

Q: Did you always want to be a mastering engineer?

You could say that; I took an interest in it pretty much as soon as I knew what mastering actually entailed, although by that time I was already in my early twenties. It's the first thing I ever did professionally, inasmuch as getting paid for it properly. I did work briefly in some smallish recording studios and did a few independent bands very early on, but it wasn't long before I grew tired of that whole thing and had the opportunity to try my hand at mastering, and it was a good fit. Prior to that I spent a couple of years working as a producer, which I loved and was great fun; but as with being an artist, a successful career as a producer requires a great deal of serendipity as well as talent, and even a successful producer can lose it all if they fall out of fashion—and I'm too pragmatic to base my career on something so ephemeral. I figured with mastering I could put in the hours, develop my skills, and work my way up.

Q: How did you become aware of a mastering process in the first place?

Well, it happened because of a couple of albums which I had produced. The first thing I did after leaving university was I started working in a little independent studio for a record label run by a guy who imagined himself as not only a great entrepreneur, but as a great talent spotter. As far as I was concerned, he was a guy who let me play in his studio, and he paid me the princely sum of 20 pounds a week, plus as many cigarettes as I could smoke. I produced sessions by these artists that he "discovered" and had a whale of a time getting into the process of producing, arranging, and programming, and all of that stuff. After a couple of years, I realized that it wasn't really going to get me anywhere. I left and moved to the other side of the country, and it just so happened that I was just up the road from the company that was handling the vinyl pressings for these two albums that I'd been working on, so I offered to liaise with the company over the test pressings and make sure they were okay. Sure enough, these test pressings got back, and I was somewhat underwhelmed by the way they sounded at home. It transpired, to cut a long story short, the reason for that was that I had elected *not* to have them EQ'd before cutting.

We'd been given a form to fill in, and there was one box that said, "Would you like someone to EQ this before it's cut?" I don't think they used the word mastering in that form, but it was just the idea of some-body that I hadn't even met making arbitrary decisions about how my things should sound that didn't really appeal, so I didn't let that happen. In other words, I checked the, "No mastering please" box, because like every young producer I was absolutely convinced that my mixes sounded absolutely fucking brilliant . . .

It's an interesting one, this. It kind of cuts to the chase of people's mis-understandings about mastering, really, because I thought my ears worked perfectly well and that I was able to be objective; however, there were various factors that were skewing that objectivity. I had taken my mixes home, I mean in those days I was taking them home on cassettes. They might even have been Dolby encoded on the tape, and I was probably playing them back with the Dolby off, so I had that ultra-compressed top end coming back at me. Then, of course, my turntable wasn't the greatest, but then again when I was listening to my favourite albums I wasn't really judging them as harshly as you would judge your own work, so I wasn't mindful of the shortcomings of my playback setup. The combination of those two things meant that the TPs just sounded very, very underwhelm-ing compared to what I thought my mixes sounded like.

Now the company that had brokered the pressings was SRT, and I called them up to explain my concerns and was invited to their mastering studio to listen to the test pressings there and A/B them with the DAT master I had supplied. The upshot of this was that—under studio conditions—I was forced to accept that in fact the vinyl pressings were a very good repre-sentation of the flat mixes. The engineer said something like, "Well, your problem is, mate, you ticked this box saying you didn't want it to be EQ'd, and if you'd actually allowed us to do our thing with it, then it would have sounded like this . . ." and then he quickly set about making a few altera-tions to the EQ.

It was kind of an epiphany moment, really; I suddenly realized how far wide of the mark I had been in my appraisal of what I'd actually done, and then also discovered that there is this whole other process, whereby some-body would actually provide some objectivity, if it had been missing, and sort all these issues out. We then chatted for ages, during which I somehow blagged myself a trial run working in that very studio, and was still there 10 years later.

It was really just a thing of having been on the receiving end—on the customer side of the fence. Being slapped about the face with the realization that mastering was what I really needed in that situation, and it would have made the whole difference. I also saw many facets of the job that suited me personally—I love so many different kinds of music, and in mastering you get that variety, so you never get bored, and you can make your contribution in a short time frame. Just prior to this I had been recording demos for bands and losing the will to live—up till 4 a.m. waiting for a bass player to get something right and thinking, "If only he'd go HOME I could nail this in one take. . . ." The whole process of having to sit there in a recording studio while nothing happens for ages and ages and ages could be neatly side-stepped if I got into a job where I could actually be given something that was pretty much finished. I thought, "This is a great job. I really ought to get into this."

Q: So you like the process, and you like sort of where it was in the production process as well—it suited you?

Yes. I mean the idea of being able to make a contribution without getting messed around. The fact the musicians didn't even have to be there—it didn't matter if they were late because they couldn't get out of bed. Also, being able to work on folk music in the afternoon, and heavy metal the following morning, then country, and then jazz, because I love loads of different types of music anyway. So to not get stuck with one particular style for like a month while somebody works on their album seemed very appealing.

Q: Yeah, I know exactly what you're talking about, that makes sense. So then, if I have you right, you cut your teeth cutting vinyl, basically?

No, not at all. I didn't cut vinyl till a lot later. But these days nearly everything goes through some kind of mastering process before the vinyl is cut, and we had skipped that process. The result was inferior sounding vinyl which then sounded even worse on my inferior deck at home—and that really shone a light on the importance as well as the creative scope of mastering as a whole.

Q: As a mastering engineer, as you're making adjustments, how comfortable do you feel intervening in the balance of a mix, let's say, without feeling like, "I'm at the point now where I should go back and communicate with the client." In other words, one of the things that's sort of a grey area is what the limits of the mastering process are in terms of "What's artistically acceptable?"

That very much depends on the client and what they expect of you, and there's huge variation in that. Some clients come in with great mixes, in which they rightly have great confidence, and they don't want something coming back to them that sounds significantly different—maybe a little level balancing track-to-track and some very subtle EQ tweaks, but they want objective confirmation that all is okay, as much as anything. They're trusting me to do nothing, if nothing is the right course of action. Then there are other clients who if you don't make a huge change to it, then they don't feel as though they've got their money's worth out of you. So it depends on them, really. You have to make a judgement about how they're going to feel about that kind of thing, and communication is of course key here.

Some clients want me to let them know if there's anything about the mix that I think they should change, while others are really not interested in my opinion and just want it mastered. That said, with so many productions now created in the box, a recall and mix tweak is often very simple to do, so I'm far more inclined these days to suggest a tweak if I think it will benefit the end product. Better to do that than to try and kludge it in the mastering.

But the limits of the mastering process are very much bound up in the mix itself. Sometimes a mix has been done very well despite very poor monitoring conditions, and in that case you might be able to apply large amounts of EQ to correct the shape, and something that sounded very wide of the mark suddenly sounds great. If every aspect of the mix sounds better, then there's no problem. But if you have a situation where the EQ you want to apply only makes most of the instruments sound better but also brings out something unwanted at the same time, then you really need a dialogue—either to adjust the mix or figure out where the compromise has to sit.

Q: When you do communicate, do you find yourself regularly communicating with, say, the mix engineer, about balances, say, before the mastering process? Or do you find that you get mixes which are generally in the ballpark to begin with, given what you are doing?

The vast majority of them are in the ballpark. It's kind of rare to get something that really needs a mix tweak. Having said that, I can be quite a perfectionist, and if something is a little bit wrong, it'll bother me. The thing is that depending on the client again, some people don't want you calling them back and throwing problems at them. Some people will deliver something to you, and they've sent it off for mastering because as far as they're concerned it's done, it's finished, it's ready. For you to then get back to them with a list of criticisms, which it could be taken that way, it might not go down very well. There are other people for whom, if you went ahead and mastered it knowing that you had those reservations without going back to them, then they'd be really pissed off that you've done it without actually telling them what your reservations were. It's about them really, rather than about me. I've got to figure out what they are expecting from me.

Q: I suppose that's just experience working with people, and knowing what the difference is?

Yeah, I guess so. You can go backwards and forwards forever trying to get a mix to sound like something else. Sometimes with certain situations, you just have to accept that's what they're looking for, and then still figure out what the trajectory is for it as well, because I quite often find that a mix has its own particular trajectory: it might not sound like it's the end result yet, but you can tell where it's trying to go. You've just got to help it in that direction rather than trying to steer it off somewhere else.

Q: So then you do cut vinyl, right? Can you speak to differences in the process when you're working for vinyl?

Yes, I do. Only in the overall sense that I think that any changes that you might make that will help a digital master sound better will also make its vinyl equivalent sound better. I don't think that there's anything that I would do differently, unless, and there is only the one caveat to this, and that is those scenarios where your client is insisting that you make your digital master louder than you would like. In that scenario you have at least some grounds to go back and say, "I think we should do the vinyl master with a bit more of a dynamic range than we've done the digital one." We find that's increasingly rare. I think very few of our clients these days want stuff to be crushed.

Q: Is that because of education about dynamic range? You know, the whole loudness war thing, obviously?

Yeah, I think so. Plus we now have loudness normalization on playback platforms such as Spotify, YouTube, and Apple Music, so people are cottoning on to the fact that the actual loudness of a music file does not often correlate with the volume at which it is played back. In addition to this, we're now attracting a higher proportion of independent artists operating in genres where sound quality is more important to them than loudness, and that's just to do with the growth of our client base via word of mouth.

It's not totally over yet; you only have to listen to the UK Top 40 [at the time of writing] to find that people are still creating appalling-sounding, horribly crushed masters, but things are moving in the right direction, and people are having hits now with stuff that is mastered with more dynamics, and you know what? It sounds just as loud on the radio! And better, too . . .

Q: When you start a project and you're going to, you know, you have the premasters, do you vet them beforehand? Do you listen to them and give feedback regularly beforehand? Or do you take the brief and then sit down to work?

What we do is we make it pretty clear to our clients that they can get feedback from us if they want it. What we try and do is to get a situation where we know whether or not they are going to want us getting back to them and saying, "Yeah, are you sure about that kick drum?" Or you know, we kind of know whether they want to hear from us on it or not. If somebody's asking for feedback, then we will give them quite detailed feedback.

Q: Do you find that people ask for feedback?

Among new clients it just tends to be those that are less experienced, or who are aware of limitations in their monitoring.

Then again there are plenty of our clients who are non-producing artists, or are small record labels or artist managers—and they wouldn't necessarily understand a word of it if we did give feedback!

With longer term clients, particularly the mix engineers and producers, after a while, as the mastering engineer of choice, you sense very much that you're considered to be part of the team. This is a great situation to be in because you know that you can be straight with people about any misgivings without giving offence, and that they will be straight with you.

But as I say, there are some people who don't want to know. They're not going to want to have their work critiqued because that's not what they're paying me to do. They're paying me to master it, not criticize it. So yeah, sorry to be vague about it, but it really, really varies, that one.

Q: So then once you have the material in, what is it that you do then? Do you have a routine or a particular process?

Yeah, well, if it's an album or an EP, I will spend a fair amount of time listening to the tracks flat before I reach for anything, just to find my way into the sound world that they've created. I also look for the common threads from one track to another, to see how much of a discrepancy there is between the tracks, and if there is a discrepancy, where is the centre ground. Then if I've got the clients there with me and there is a discrepancy, that a couple of tracks are much quieter than the others, I can then say to them with quite specific reference to what we're listening to, you know, "How do you notice these differences, and if so, what do you prefer? Do you prefer the brighter ones?" Then they can get an idea of where they're going to want to go with it. It's not often that there's a massive discrepancy, but it does happen from time to time. I think step one is to just have a good listen.

Q: Do you tend to work with a particular chain or chains in mind and, you know, material coming through that's in a particular vein? You know you're going to work with these particular technologies, or are you completely open?

I try to be completely open, but as a starting point I do have a default signal path. If, during the course of the beginning of the session, I discovered that there are certain, let's say, analog units in that path that I'm simply not going to use, then I'll just take them out, just to keep the path as clean as possible. I usually start with them all kind of in so that I can experiment with them at the beginning of the session and figure out what will work best, particularly if there are two tools that do a similar job, and you just don't know which one is going to do it right for that particular track until you've given it a go.

Q: Once you've set up, once you've found roughly the chain that you like, what are the things that you're listening for? As a mastering engineer, what are you listening for in the material that you're going to then know that, "When I've done these things, I'm now delivering a master"?

It's one of those things that's quite hard to put into words. I think, in a way, I'm starting off just trying to be aware of things that are bugging me; I guess that's step one. Is there a certain [frequency area] which might happen if there are resonances in the room where it was recorded, or certain areas where there's a little too much energy in a certain frequency range? That's the kind of thing that is kind of remedial, if you'd like, and it presents to me as an annoyance while I'm listening—something that just keeps popping up and bugging me. Things like that would probably be the first things that I would actually attempt to address. Those things where I might use a notch digital EQ, or a dynamic EQ, or something like that to sort of tuck things in before I can then start looking at what there is around that area that can be enhanced. That's the thing: if there's something that is really poking out in a certain range, it can then mask other stuff. If you can actually tuck it in, all of a sudden you think, "Well, actually, all that stuff around it, it's kind of there to be enhanced now. We can start bringing that range forward and give a bit more definition to that." I guess that's something that's fairly common—I'll be looking for things to notch out digitally, and then when I want to enhance the things around that range, I'm more likely to use an analog EQ to lift them up again.

Q: Do you have any favourite tools in that regard? I know it's going to depend on the material.

Yeah, when it comes to notching stuff out, I've got the TC Electronics System 6000, and I've got the Massenburg EQ algorithm on there. That's what I'll usually use for notching down problem frequencies. Occasionally, there might be something that I'll be doing in the mid/side domain, in which case I'll use one of the other engines in the TC for that. Then, when it comes to the more tonal shaping side of things, it's mainly the Manley Massive Passive or the Prism MEA2. I kind of bounce between them, really, depending on what the range is, and probably everybody says the same thing, really. The MEA2 is really nice and transparent, and it's nice for adding air at the top. It can be quite tight for adding a little bit of bass. The Manley is more colourful and a bit less neutral—great for warmth and character.

Q: When you were talking about your default chain, would you go through what that is?

Right; so the default chain consists of a Weiss DS1. That's a de-esser, so that's the first thing that we hit. Then going into the TC Electronics System 6000, I have the Massenburg EQ. Then one of the utility type MD4 Engines, which is also used if I need to sort out the gain before it then gets into the analog domain. Whereupon it gets to a Manley Massive Passive followed by a Prism MEA2. There's also a Prism MLA2, which I find that I use very, very infrequently these days. I've got a Dolby Spectral Processor, which is my ace in the hole, that's the thing that I've got, which I don't use it very often, but there are not many of them around, and there are certain things where it's a real mix saver, I find. It's great for just providing a little bit of detail, and also for those "inky black" digital recordings that don't seem to have any ambience in them at all, it's great for just sorting that out. Usually it's in bypass, but occasionally it's really handy. I've also got a Thermionic Culture's Culture Vulture, and that's next. Then, the last analog bit usually in the chain is a Rupert Neve-designed Portico 2 Master Buss Processor. It's a really nice compressor; I like it a lot.

So after that, we're back into the digital domain. There's a final engine in the TC6000. Then we are back into SADiE, which is sort of my core workstation.

Q: You were talking about notching out resonances if there are any, and that allows you to work around the area. What happens next?

I guess it's EQ first and then dynamics. I don't think you can address the dynamics until the EQ is sorted, or at least very close. So, depending on the genre of course but assuming it's something with a backbeat, I'll start thinking about punch and gel. I don't like multiband compression much, I often find it makes things too dense and bland—but the exception can be bass compression, which I might use to tighten the bottom end and give it the requisite poke. Then I start to think about overall compression for gel and punch often not needed because the mix engineer has often taken care of that.

It's not a conscious thing: I recently started to think about my work process in terms of spectral dynamics. That is to say, in each frequency range there are loud sounds and quiet sounds, and I'm dealing with the relationships between them. Let's say, for instance, that in a given track the chorus is much louder than the verse, but the drummer switches from a high-hat, which is loud, to ride cymbals, which are quiet. In that scenario, I might want to brighten the chorus to compensate, whilst at the same time compressing a bit if the overall level change is too great. Or maybe the top end of the snare is dominating the upper mids to a degree that the rhythm guitar isn't quite working. There's 8s on the hat and the snare on 2 and 4, all of which are dominating so much that the rhythm sounds lumpy and boring, whilst the 16th note syncopation in the guitar part is being overshadowed. Well, given that the edgy choppy area on the guitar might be around 4 kHz, as is the top of the snare, I could choose to use a de-esser to hold back those snare peaks. The body of the snare still comes through as strong, but now the toppy peaks are held back a bit, allowing the rhythm guitar to come through (maybe with an EQ lift), and this suddenly gives the track back its groove.

So I'm looking to get that moment where I feel as though everything that's been laid down, everything that's been put into the mix, is having the desired effect. I guess that's what I'm always looking for: what's the desired effect? I have this thing where I'll listen, and I'll listen, and I'll be looking at one thing, and then another thing. Eventually I'll reach a point where I think, "You know what? I think this is as good as it gets." And that's when I put it down, which is terrible really, because I think what customers really want to see is sometimes that kind of showboating thing of, "Yeah, man, this sounds f*cking amazing!" That kind of thing. They want to be given this huge confidence boost. I'm a little bit more of a chin-stroker; I'm just, kind of sat there going, "Hmm, yeah, you know, that'll do it." Then when I listen to it the next day I might think, "That's really, really good." But when I'm in the moment, I'm thinking so hard and just trying to achieve the optimum effect from what's there. Sometimes it just doesn't quite get there. It might be as good as it's possible to get it, but in the moment I'm thinking, "Oh well, it's going to have to do," because I'm always looking for more. That's just my perfectionism coming through. Sometimes I wish I was a better actor!

Q: Now we're going to sort of shift a tiny bit then to the very end of the chain, and talk about meters and metering, if you don't mind. Are there any meters that you typically rely on, and is there a reason for it, or is it just that they're what you use?

It's a bit of both. The two things that I'm using most of the time is, I've got a pair of standard analog VU meters which I've got half an eye on. The other thing that I use is we have a Cedar Cambridge workstation, because I used to do a lot more restoration than we do these days. I've always used Cedar's equipment for things like de-clicking and that kind of thing. The Cambridge system is great because it's got azimuth correction and stuff which is really, really handy. Nobody wants to pay for that kind of level of precision for restoration anymore, but we got it. It's also got a phase display on it. It's also got a really sophisticated spectrum analyser. Then there is a ballistically slowed down RMS meter on it. That really is what I'm using to keep an eye on loudness. One of the things I like about it is that it's on a different screen. I know there are various plugins you can get now, but my main workstation monitor is cluttered enough as it is, so the fact that Cambridge is running on a different PC, I have a separate monitor for it, so it's always there. I find it very reliable for keeping track of dynamics and loudness, and that kind of thing. Whenever I do try out other plugins, it seems pretty consistent in terms of what it's telling me.

Q: Do you find yourself pegging to particular values on these meters? In other words, have you ever worked on material and your ears deceived you, and you used the visual information? Or is it just purely confirming what you hear?

I think the true honest answer is always going to be somewhere between those two, to a degree. But no, for the most part, it's just to confirm what I'm hearing.

I guess what I'm saying is this: if you're watching the meters all the time, I don't think anybody in all honesty would be able to say that they're working completely off their ears when they're looking at visual cues as well; it's always going to have a degree of influence on how you perceive what you're doing. That's the way we process information coming in through our various different senses. You amalgamate it internally, somehow.

There was one situation where my meters were telling me something completely different to what my ears were telling me. That was just in a bizarre scenario where I was working on a 96 kHz recording. My compressor was behaving really weirdly, and my meters were showing me some really odd stuff [that was] going on. I was a little bit baffled, until I eventually used a spectrogram, which was a Cedar Plugin called Retouch, which actually demonstrated to me that the recording that I was working on basically had a spectral mirror image of itself centred around 24 kHz. Basically, everything that was happening in the low end was also happening up at around 40 kHz. There was a ridiculous amount of energy in the very high top. It was all completely inaudible, but obviously the meters were showing me it, and my compressor was reacting to signals that I couldn't hear, even in the analog domain. It was good to have that information to hand, because I was then able to put in a low pass filter, which then at least caused my compressors to start acting like compressors again. That was very, very weird, and I've come across it a couple of times since but not been able to determine the cause (although I think in the first instance it was a fault in the guy's sound card).

Q: Let's say that you do find an imbalance in the stereo image—will you go to rectify that? Or is that just the grounds to pick up the phone and talk to the mix engineer? Or are there things you do?

I do have a policy on that, which is that if we're just talking about a left/right imbalance—let's say, if we're assuming that the arrangement is like a fairly basic, straightforward pop arrangement, bass, drums, guitar, lead vocal—if it's leaning heavily to one side, then I would look at things like the snare drum and the vocal. If they're still in the middle, then I'm going to allow that track to lean, because I'm going to take the view that most likely, the vocal and the snare drum were intended to be in the middle, because that's where they are, dead centre. If the guitars are pulling it to one side, then they just decided to have the guitars pulling it to one side. If the customer was sat with me in the room, which is quite often the case, then I'll talk to them about it. If there's something that's normally in the middle and it's still in the middle, then I'm going to assume that that's the way it is supposed to be, rather than me trying to change the balance and therefore shifting things over to one side. Really, the idea that the left and right channels should be equal loudness at all times is a pretty arbitrary notion. I think it's more important that your vocal is in the middle, unless you have reason to believe that that wasn't what they wanted to do. There might be certain cases where that might be what you would think, so yeah, I would pick up the phone.

There are other enhancements to the stereo image that can happen. Sometimes I use mid/side EQ, which like any tool can be dangerous, but I sometimes will EQ the side image in order to improve definition of elements that are more widely panned—and that can also result in a more immersive stereo image.

Q: I saw that a member of The Kinks was very happy with your work. It's advertised on your website, which is nice. So I was hoping you'd be willing maybe to talk about some of the differences in your workflow, or in similarities as well with regards to remastering. It seems to be a separate thinking form of practice within the broader mastering process. And there's really not a lot of information out there about why it's done, how it's done, how it differs? So it's a very open-ended question, but I'm wondering if you'd be willing to discuss that process?

Let's see what comes to mind. I think when it comes to why it's done, there are a couple of things. I think the most obvious one is that anything that's been mastered, any catalogue stuff that was mastered in the very early days of digital, the A/D converters back then were pretty horrible compared to what we have now. That reason in itself is a good enough reason to do that process again now or at a later point. Of course, the other thing is that a lot of much loved recordings were not necessarily treated with that same love when they were actually put out the first time around. A lot of things were knocked off fairly quickly, and so it's nice to maybe be able to spend a little bit more time on sorting out things that couldn't be sorted out at the time. Then you've got to balance that with the fact that what went out at the time is what is loved, so you don't want to change it into something else. If there are certain things that you can do, I mean I was talking about notching out resonances before. You have to make a judgement about not going too far with these things. But if a particular recording was compromised by a particular frequency that's sticking out, and you can address that now because the EQ that we have now is far more accurate than what was available at the time, then it makes sense to do that, as long as people don't end up putting it on and thinking, "This doesn't sound like the way I remember it?"

The thing is that people have rose-tinted memories. You could give them a remaster that sounds identical to the original, and they might be disappointed, whereas sometimes you can give them something that sounds different—enhanced in some way—in order to give them the feeling that it sounds just like they remember.

That said, it needs to be done with taste. I've stopped buying remasters, myself. As a consumer, I'm absolutely appalled with some of the stuff that's come out in the last 10 years. Prior to that, I think it was a really valid thing.

I don't do as much catalogue as I used to, but recently completed a box set of albums by Pentangle. This was interesting to me because of the range of production parts available to me to work from; I had analog tapes—generally production copies from the period, or flat copies done a few years later—the best and most original sources that could be found, anyway. Then there was the vinyl, a complete set of first edition pressings—for reference. Finally, the CDs as these albums had all been remastered already a few years ago, it made sense for me to have those digital versions as a reference, too.

The tapes sounded okay. They were nearly all Dolby tapes, so the top end was a touch murky and brittle, but the tapes were a good starting point, and in the absence of anything else, I'd probably have mastered the whole lot quite closely to those tapes. The vinyl, however, sounded very different, generally much warmer and with a lot more midrange, almost dull compared to the tapes, although there was something about the top end that I liked. Then there was the CDs—which represented the sound for this material that most consumers would have been used to. The mastering on those had gone in the opposite direction, and they were generally brighter than the tapes. Having these different reference points was really useful in creating the new masters, and the vinyl in particular gave me licence to go for a far warmer sound than the band has had in the digital domain before now.

Q: Do you think that has to do with an attitude within remastering, where sometimes they're advertised, at least, as, "Oh we've taken the older masters and we've updated them"?

You have to bear in mind that a lot of stuff comes out, and it's got a badge on it saying it's been remastered. That doesn't actually mean necessarily that it's been remastered as we professionals would understand that term. I know for a fact that there's albums that I've worked on, that you'd see in the shops, and there's a big silver sticker on it saying, "remastered from the original analog tapes." I'm like, "No, it wasn't." Somebody gave me a CD and said, "Oh, can you recompile this and turn it up a bit please?" They only paid me £300, and I wouldn't have been able to do a proper remaster for that budget. Then no one gave me the original tapes anyway. I mean that has happened, and I think it has happened through the industry massively, because the mastering is down to some production manager and the record company, and then the marketing is down to somebody else who then goes, "Oh well, this is an old album. We need to put one of those silver stickers on, saying it's been remastered from the original tapes." They don't actually check to see if that was actually done. There have been all sorts of travesties that have gone on. I think even within the era where stuff has been done properly, I would question the taste that's been applied, because people have done stuff that sounded great a few years ago, and now they've made it too loud and too bright. Whether they think they're trying to please a certain audience, or whether they're just jaded by the fact that they were working on something very current the day before, I don't know. You go online now and look at forums of people that are consumers of catalogue material, and you'll find loads of people understandably fed up with the fact that remasters frequently sound nothing like as good as the originals did, which is really what it's about.

Q: That's another thing we've noticed, that there is a great deal of cynicism towards the notion of remastering now.

Very much so; it's understandable. I think that a really cynical person would probably slate it as being a marketing ploy from the get-go, regardless of what it sounds like. I think there have been enough really awfully sounding remasters out there that have been well enough known to give that cynicism an extra weight, I think, which is a shame.

Q: Do you get issues where the original tapes or original digital transfers for reissues are lost in label catalogues?

Well, that kind of thing happens a lot. The question is, are these assets actually lost, or just not easy to find? So it depends then on your relationship with whoever's actually booking the work in with you. Certain people, once they appreciate that your opinion is worth something, are prepared to move things around for you. Until you're in that position, then there's the risk that you're going back to people saying, "Look, these sources you're sending me are rubbish," and they eventually start thinking, "I don't like sending stuff to that guy, because he just gets back to me telling me it's all rubbish. I'm going to send it to this bloke who just compiles it without complaining." So you have to be put into this awful position of having to be pragmatic.

Q: Are you willing to speak about some alternative ways of achieving, quote unquote, "loudness," other than, say, compression or limiting?

Oh absolutely, yeah. Distort it! [Laughs] The thing is that there are various different limiters that you can get, but at the end of the day they all do pretty much the same thing. There comes a point where having gain reduction applied in any form is going to take stuff away that you don't want to lose.

Well, I guess, just the energy contained within a moment where maybe a snare drum or a kick drum is happening—in order to fit that into the space that you're fitting it into, all the limiter is ever going to do is to turn the volume down for a moment to allow it to fit. That has a certain evening, softening, effect on everything. In some instances, you can get around that by allowing a signal to clip; you can use soft clipping. I occasionally use a soft clip on the TC6000, which allows you to manage peaks without gain reduction being applied. You can even just hard clip the A/D converters. You can even hard clip the signal after it's been loaded back into your workstation. There are situations whereby, despite the grotesque effect that has on the waveform, it actually sounds cleaner, and it sounds like it has more energy to it, than trying to achieve the same amount of loudness using purely limiting. Though there might be a case for limiting it up to a certain point, and then clipping it a little bit afterwards. Now I don't advocate it, but it's something that I have occasionally done to good effect in those situations where my client is insisting that I compete with something. I don't think it ever sounds better than if you haven't had to do that in the first place, but it can certainly sound, to all intents and purposes, no worse. Particularly if it's something that's got a very strong beat and not very much in terms of sustain information—you can get away with clipping stuff in that case. I tend to never want to go too far with that kind of thing. I hear stuff that's been commercially released and very successful that sounds horribly distorted to me. I just wonder why people have allowed it to go through.

Then again, there is technology available now that manages to increase average level massively without distortion artefacts. Everything super loud, smooth, fat, dense, and incredibly boring. Every bit of space taken up with sound. That kind of thing only sounds good maybe on a cell phone— I can't stand it. Music needs space. A great recording played on decent monitors sounds visceral, alive. These super-hot masters with a dynamic range of 5 dB or less just sound dead to me.

Q: Do you find yourself working differently for different formats? Do you have a different ideal end state or ideal master when you're working for something that you know is just going to get a non-physical release, versus something that is going to get a physical release? Or do you tend to generate a master, and then that's used for the different formats.

It's normally the latter. We go for an ideal master, which has a little headroom to avoid inter-sample clips, and in the case of MfiT masters [Mastered for iTunes] there's an extra quality control check, and sometimes a further level adjustment is required.

Q: You mentioned that you check inter-sample clips, and that's a very specialized bit of information. Can you talk about that? Do you meter for that and adjust? What do you use for that?

Yeah, there are the Apple Tools for MfiT, plus for real-time checks we've got the Sonnox "Fraunhofer Pro-Codec" plugin, which does real time encoding either to mp3 or AAC Codecs, whichever you want to test for. We can run something through there and get an idea of how often and how extreme or audible the inter-sample clipping would be.

Q: Are you noticing any broader trends within the industry at large? Or you know, the sort of sayings: "Mastering houses, they're tending to close down"? Is it becoming more of a boutique affair?

There's definitely some evidence of that, I think the boutique places, of which I guess we're an example of one, and we certainly weren't the first.

I mean in the UK, I guess when people like Kevin Metcalfe left The Town-house and started up The Sound Masters, he was probably one of the first people to move away from one of the big studios and set up independently with a facility that was equal in terms of its technical prowess and capabilities. Then, in their wake, quite a few people have done likewise. You do have a larger number of options, but all with studios and staff that are sort of world-class quality. Meanwhile, sadly, the big places have been struggling. It's just very hard to make money out of this when you've got to pay for 24-hour maintenance staff, security, canteens, and all of that ancillary stuff that goes with a big studio complex.

It's just that the money's not there to support that, I think, essentially. When people are allocating their budgets for studio services, they're expecting to pay much less than they used to. They're also getting stuff delivered to them from much smaller places all the time anyway, so they are able to do things much more cheaply. So then the onus is on people like us to try and maintain the standards, but with lower margins and less money floating around, basically. Then the other thing of course that is happening is you get much more of these smaller "mastering services" popping up, like people online that have a computer at home and a couple of plugins. As long as they've got a nice-looking interface on their website, they can call themselves a mastering studio. For a while they seemed to be popping up all the time. The distressing trend that comes off the back of all that is that it corrupts people's notions of what mastering actually is. I once started a mastering session with a new client that I'd been recommended to, and they said "we're a little bit nervous about this whole thing, because on our last album we had it mastered and we didn't really like it. We don't think we like mastering, because we tried it before, and it made our music sound worse." This is a flabbergasting statement, really, but when you think about it, with so many online mastering setups run by self-trained semi-pros, and so much misinformation around, it begins to make sense. I asked "Well why did you accept it?" . . . and for whatever reason they thought they had to, but the thing was that whoever had done this mastering for them before was not someone I was aware of, and it made me realize that people are getting bad mastering done and not even realizing it. They're sending their music to people who haven't got the foggiest idea about what they're doing. Then they come to you, and you end up starting off the session on the back foot having to demonstrate to people that you're not going to do that to their stuff. It's a bit of a weird one.

Although, I think it's possible that the "writing is on the wall" for a lot of those people, because now there's LandR.com.

The other thing that I used to have to tackle with people is that situation where people say, "Well, how much of a percentage difference is it going to make?" But that's the thing now: so many more people are making music, and many of them have never read the credits on an album sleeve in their life. Somebody has told them that they need something mastered, but they have no idea what that's for and what it's all about. Somebody has said, "Well, it will make your tracks sound better." Then they're on the phone to us now, and we've been talking about the price. They're saying,

"Well, okay, if I give you that much money, how many percent better is it going to sound afterwards?" I have to say to them, "Well, I can't answer that question. If it sounds really good, then I'm not going to do anything to it at all. You still have to pay me though." It's for my time to sit there and make that educated decision to not do anything, whereas you could've taken it to some guy for a fiver and ruined it.

Thankfully, these days, far more of our enquiries are from people who know exactly what they're getting. It does make life a lot easier.

Q: Are you finding that things are moving towards EPs these days, less album work? It seems to be very much "drip feed" material for the Internet fodder, rather than cohesive albums.

We're very fortunate in that we do get a lot of album work, which is really fulfilling as well as being best for business. Creatively you can get into a zone with an album and get a lot more material mastered in a shorter amount of time, which is better for us and the clients. But yes, there is definitely a trend towards singles and EPs. I don't mind, I think EPs are fine. If the EP ends up becoming the primary format in a few years' time, I don't think it will be disastrous. I do like albums, but I think EPs have something of their own to offer. I guess there is a little bit of a shift going on there.

On the other hand, we get quite a lot of people wanting to come and master one track. That's a real pain, because usually one of the first things that you're doing, as I was saying at the beginning of this, really, I want to sit there and have a listen to the tracks and find a common thread, you know.

When a track comes in, in complete isolation from anything else, it's a lot harder to actually figure out quite what your role is, sometimes. Unless you know something just obviously needs a whole lot of stuff doing to it. It's so much less economical as well. It's like with our online prices: we have to charge a little bit more for the first track in each session if people are paying by the track, and then we lose out, anyway. It takes so long to sit there, get yourself in the flow, and just master one song, and then that's it. Then there's the admin involved—office time, etc., just for a single track invoice.

Q: Does it tend to depend on genre?

Yes, I would agree with that, but I think also our client base is more geared towards traditional and album-based type artists. We haven't sought for that to happen, but it has just seemed to have naturally gone that way.

Q: Do you find yourself doing particular things to project studio recordings that you don't do to those that have been through a more professional production process?

Well, only inasmuch as you're more likely to have to get a little bit more creative to get the right results out of the mix. Obviously the better, more experienced a recording and a mix is, the less likely we are to have to do anything to it. With the project stuff, it's a lot more of a mixed bag.

Occasionally, something will come along that's been done in a really cheap studio somewhere, or on a home studio, that just sounds absolutely amazing.

You know, recordings can have quirks that are charming as well. Sometimes you don't want to iron everything out because sometimes those things are what make a recording interesting. Obviously, with a project studio, they're more likely to have gotten the bass wrong or not gotten the best out of the bottom end, because that's the thing that's hardest to monitor in a project studio. In that situation, we can always work with stems, but it's not happening as much as I thought it would—the stem thing.

I'm "heading the stems off at the pass" quite a lot, because—when people come to us and they're asking us about mastering from stems—because I don't want to get into a situation where we're having to mix it for them. So I'll have a conversation with them whereby I'll make it clear that it needs to be a mix that they're happy with first, so that it can be a mastering session and not a mixing session. Through the course of those conversations, I quite often discover that actually what these people have is a mix that they know isn't actually ready yet, and that they need to have it finished before they have it mastered.

The thing is, on the one hand, I'm saying to people, "Look, if your mix isn't finished, let's not go ahead, because I don't want to be running up a massive bill that you don't want to pay." You need to wear two different hats. Now, I have done some stem jobs where I have had to start off with a little bit of a "mixing hat" on and have come up with some really great results, but for the most part, I don't want to have to go there, because mastering is not mixing.

It's simpler now, because at Fluid we have set up a mixing service that is quite clearly, categorically, a different service. Better to have something tweaked in there and prepped for mastering as a stereo file rather than go straight to mastering with stems. Having to put a mixing hat on when I'm supposed to be mastering—to me, it prevents me from doing my mastering job properly.

Q: If you had to define audio mastering then, how would you define it as a sort of art form, as a craft?

I've got different ways of describing it every week. I've been doing this for 25 years, and I still kind of take the view that it's a little bit like some guy has created this wonderful sculpture, and our job is to position it and light it right so that everyone can see it from the best possible perspective. We're not changing the sculpture, we're not changing the form of it, but we're just allowing everyone to see it in the best light. That's one way of putting it.

Someone was asking me recently about the difference between the skills that you need for mixing versus for mastering, and I was saying that I think there is a problem with people that try and do both. There are loads of recording and mixing studios now that will offer mastering as an add on. You know, "We'll mix your album for this, but for an extra whatever, we'll master it too." It's like, "I don't know how you can actually master stuff when you're mixing all the time." It's such a different mindset, really.

I guess the differences are that when you are essentially a mixing person, or in mixing mode, when you're listening to something, you're thinking about the relative level of all the instruments, and the relative tones of the instruments to each other. You might even be thinking about the reverb on the snare drum and all that kind of stuff as well, but it's really about the relativity of things within that. As a mastering engineer, you simply do not have access to any of that stuff. You're having to look at the relationships between different frequency ranges, or you're looking at the dynamic activity within a particular frequency range. It's a little bit like having your hands tied behind your back to begin with. You need to do that for a while; you need to be in a position where you are working with your hands behind your back for quite a while before you can figure it out. You need to drive the car with your hands behind your back for a quite a while, before you can figure out how to steer it with your nose.

Eventually, you find that there are all these things that you can do, and you'll start the session, and a number of times it's happened, I'm starting a session with a band, and they've not done it before, somebody says, "Can you turn the rhythm guitar up in this song." I'm like, "Well, no, I can't." And they go, "Oh. . . ." So then you do a few little tweaks and blah blah blah blah, and they go, "Hang on a minute, you told me you couldn't do that?" It's like, "Well, I haven't done that, but you feel as though I have." Because you just find a way of getting the perspective to work, but without doing any of the things that a mix engineer would do. It's the same; I was about to say that you're using a different set of tools, but of course you're not. You're using exactly the same tools, but you just don't have access, you can't get under the bonnet. I'm throwing metaphors all over the place.

JP Braddock

John Paul "JP" Braddock is a highly experienced audio engineer focused in both analog and digital domain mastering. He was the former head engineer of the Rubber Biscuit Studio, where he discovered his passion for restoration mastering and spark for the creation of the British Music Archive [BMA] heritage restoration project. He delivered hundreds of albums over the last two decades spanning a wide range of genres, from work with Mercury through to Earache Records. This led to JP's development of the Full Dynamic Range High Definition [FDRHD] audiophile mastering grade delivery format at his latest facility, Formation Audio Ltd. His career also features a wide range of commercial educational experience, delivering lecturers and seminars for industry events and universities across the UK.

His not-for-profit interests also include active membership of the Music Producers Guild [MPG] and AES Mastering Group [AES:MG]

Q: Did you always want to be a mastering engineer?

No, I wanted to be an architect.

Q: How did you become a mastering engineer?

I was always into music, from an early age. Taking classical piano lessons and finding this too restrictive, as many do, I started getting involved with bands, playing any keyboards I could get my hands on and faffing around with electronics. After finishing college, I didn't want to take an architecture degree for seven years. It seemed an eternity at the time. I just continued on with my music, and which led me to eventually setting up my first studio with two other partners. I was in my early twenties at that time. From that point on the progression of engineering my own music to producing bands' albums naturally led me to being involved and enjoying the aspects of mastering projects—that final overview.

Q: How did you first become aware that there was such a thing as a mastering process?

I always knew [mastering] existed as part of the chain of production. But from the first mastering session I attended as a musician at the Townhouse, I became highly interested in the process. It just made our mixes sound so much more cohesive, brilliant! I just wanted to know: "How did they do that?" I'd been aware in the same way as the need to focus on aspects of recording and mixing. It comes from being into music or a musician—enjoying listening to music. You want to create music, and you want people to hear it at its best—make it perfect! You need to investigate how that transpires if you're actually genuinely interested in doing it.

Q: At this point in time there wouldn't be a lot of information published about how mastering makes the mixes sound better. How did you go about learning this?

Between the periods of when I was 19 to my late twenties, I did quite a lot of mixing and production on albums. I was involved in going to several mastering houses. You're viewing that as an external and questioning why certain processes are being applied. There's a parallel to your development as a mix engineer in that you start to understand what compression actually is and why and how you're using it, and it's the same way with EQ. Hearing what needs addressing at the end informs the rest of the production process next time around. Being able to hear work you're intensely intimate with after another highly experienced engineer's perspective is enlightening. I found this with production work, too, where we delivered our production multi-track for mix. Comparative critical analysis of audio in all its forms, especially with reference material, is key. If there's one thing I always say to new engineers is: "I don't have any magic skills or secrets that make me a better engineer than you; it's not the tools I use or knowledge to utilize them. All that you can learn; what is hard to master is the skill of listening. The one thing as an engineer I have over you is I've listen to music every day for probably two or three decades longer! That's a lot of understanding of average I've built up. That comprehension of what sounds 'right' comes from that understanding of average."

Q: One of the themes we're pursuing is mixing and mastering and their relationship, almost as a broader, singular metaphor process. I'm wondering if you might speak to how your knowledge of the mastering process informs the specific decisions that you make, knowing what you know about what will happen to the mix once it's a premaster.

For me, those two things are starkly different. I don't see there's a lot of crossover unless the actual mix is very simple, such as an acoustic guitar and vocal. I like mixing, but I haven't done it for over a decade and a half now. As in, I won't proactively pursue mixing work. I find mixing a very emotional process. It involves a lot of your own creative input and energies in a way I don't do in mastering. In mastering, I can be much more separated. It's a technical process of creatively changing the audio, but I'm not emotionally involved with the music production process. For me, it means I can keep my music and mastering work quite separate. It took me quite a few years to work this out.

When you think about stem mastering, you don't have that distinction; this is something I'm not necessarily a big fan of. It's a fix-it tool if needs must. When working from the session mix you get an objective opinion; simply, we're not dealing with individual instruments' balance but pulling the whole mix around to do that, thus maintaining the original mix intent, making it fit for purpose, translating the emotion of the mix to function for digestion in the wider world.

The more you know about microphones, the more you understand sound moving in space, the better you are at recording. Equally, the better you are at mixing or mastering. I think all the disciplines interact with each other from that perspective. In the simplest terms, sound and manipulation of is always a combination of time and amplitude: everything comes out of this. The more you comprehend how this fits in the better you'll become as an engineer in all aspects.

Q: Can you expand a little bit on what you just said—how to see that the more you know about microphones, the more you know how sound behaves, the production process in almost holistically related? Can you speak to that at all, anything specific?

Yeah, it's things like being able to hear phase issues. From a development point of view, one of the things that's hard to grasp is what you're actually listen for. Critical listening without lots of experience in the subtleties of phase is difficult, especially experience working within multiple microphone setups when you can hear that there's something going on in there. That "issue" is going to manifest to be much more apparent later down the line when we're more compressed at the mix. Then when we get to the mastering stage and we're reducing overall RMS aspect of it, nothing about limiting, just the general density of it. Equally, at that point, it becomes more and more present. An awareness of those interrelationships of phase, I think, are key to achieving the clarity we would expect in a "professional outcome."

That would apply within equalization in a similar way. The overuse of EQ within mixing, or "Why we are even EQing on the desk when we're recording it down?" Move the microphone, change it in its space or the source itself to get the actual tonal outcome you want. You don't have to apply anything to it. That would apply as much within the creation of a sound within the computer, too. I hear a lot of over-processing with younger engineers. The more they comprehend in terms of process; as the great Miles Davis said "I always listen to what I can leave out" the better it gets. That doesn't mean you can't kill stuff or "make in nasty" or process the hell out of something but you need to know "why" each aspect is applied. Also, the comprehension that changing tone doesn't mean to adjust the EQ. Compression is the best tonal shaper we have at our disposal; it doesn't change the phase in basic use as equalization does: it changes envelopes, hence transients, and so the tone, and so on.

Many of the great production innovators over the decades, like people I've a lot of respect for such as the The Dust Brothers, think about the creation of the melody with the sounds. If the sound's inherently in the right pocket and register in all aspects, it means when you come to mix it, it all fits together. There isn't loads of post-processing to it in the mixing stage to make all of those different tones work. The tones inherently were designed to go together. I think another great exponent of that in the dance scene was Royksopp. In a similar way, they do just amazing stuff with drums. All the transients are pulled out with the distortion, but it doesn't sound distorted because they really understand how to create the tones to work with each other. Everything again sits in its pocket, especially in the harmonics, which means you don't have to do a lot to it later on—no excessive equalization need to correct, hence less introduction of phase in the mix. In that way in, the music you end up achieving really has good density, which means no one needs to limit the hell out of something to achieve volume. The volume exists because all the mix works. The sounds worked with the creation of the melodies in the first place. It's proper art.

Q: Can you speak to other means to achieving loudness aside from the limiting?

Limiting for loudness to me is something that happens after we master it, in that increased volume for the sake of commercial loudness—not because it sounds better. Limiting is useful throughout the chain to reduce peaks where peaks arise, but I don't think about limiting in terms of loudness as part of the actual mastering process. It's something that's there that can facilitate more perceived volume if required, relative to the production outcomes: the actual volume. As in how loud we make something. Its actual density comes from really good internal dynamic and tonal balance. If you've got those two things correct, you can apply the right type of limiter, which, generally most of the time, with naturally based instrumentation, for me would be BL2 off the TC System 6000, either with or without soft clip, depending on what you're trying to achieve tonally, to get the relative outcome of what you would see in a "relative commercial scale" anywhere from −12, −8 dB below full scale. That for me is something that happens because we've got the correct density of compression, in terms of dealing with the RMS to peak, right. The tonal manipulation is right to make all those things fit. If you've got that all right, you've got the density. What you can't do is take something that isn't mixed well and achieve that outcome without damaging it in some way with the limiter. There's plenty of stuff you can do if the mix is excellent; again, because the composition is excellent, the sounds are excellent, and they all fit in the right registers together, and it's not everything jumbled up on top of one another.

So in short—it's simply achieving good internal tonal and dynamic balance. We don't need limiting for loudness to achieve this.

Another example of where engineers don't deliver that unmasked clarity is the classic in DAW-age people who've only mixed in the box, not realizing pan exists or, as I've experienced a lot, "Logic" [Apple OS based DAW] mixes, everything's coming out as a "stereo" source. Mix is all panned hard left, hard right, and centre. You've got three mono mixes sitting disparate of each other. This is inherently poor in the mono sum aspect important for many delivery applications—superficially sounds wide, but this lack of panning between LCR means by its nature it's masked. The other 98% of panning points are unused. All the transients are sitting on top of each other. Unmasking each aspect of the mix using the panorama of the whole space opens up the sound massively, in the same way register does in instruments' frequency relationships. The more unmasking, the more clarity, the less volume needed to achieve clarity between parts; hence, the mix is more dense and in the end louder after mastering, with less processing. Win-win! In saying that, some skilled mix engineers like to pan this way, but they've got the mix to work in the pocket in mono first and then pan hard for that stereo effect.

If you get all of the parts together right with people who effectively know how to mix, in mastering, we can draw all the essence of these qualities out, and all you really are compensating for is the environment it's mixed in, plus making the mix more cohesive or glued in terms of a useable dynamic for the delivery. It's that finesse of being able to make the total balance sit, so all of the detail becomes apparent and present where it needs to be, as opposed to it being slightly masked in places. You can't take something that's badly mixed and make it sound amazing. It's just not possible. You can change it a lot, so it might sound much better, but that doesn't mean it sounds as good as it could have. I suppose a lot of the people I work with are conscious of trying to improve their ability to produce and mix to achieve a better outcome. I don't see it as,

"Here's a job. Get it done. Push it out. Next one." You're building up relationships with producers and engineers to achieve better product. If we're not there to make music the best way we can, why are we doing it, really?

Q: How did you go about initially getting clients? How did you become professional as a mastering engineer?

The first thing here was I didn't actually ever call myself a mastering engineer for probably about 10 years while I was mastering—I had too much respect for my peers to feel comfortable in myself saying that. I didn't have the "years" of listening experience. Whereas, presently, modern times are such that people will call themselves anything immediately—like, "I'm a professional cyclist because I bought a pro road bike." I think that's the nature of the times.

At the time when I was getting into it engineering, digital was in its infancy, but in our first studio venture we were using a 32-track Soundscape system. [Soundscape was a hardware-enhanced DAW using the PC for a GUI and control.] We were one of the only places that had digital editing at that kind of level in the Midlands. So we weren't necessarily brilliant at recording or had a really nice live room, but what we could do is manipulate and edit mixes quite well. We ended up finalizing a lot of stuff and sending it to be mastered to reduce costs for clients before mastering—almost what you could call "stem editing" in many cases, remembering at this time it was about £200/hr to book a mastering session. We'd be editing things up, sorting out track spacing/edits and simplifying the processing a little bit.

I was also taking mastered projects back that I'd produced or engineered that I wasn't necessarily happy. I was spending quite a considerable amount of time trying to improve that to achieve a better outcome. Obviously, you do that and achieve something that's a better outcome, and that's not to be disparaging about those experienced engineers doing the masters, but the difference was they were spending six hours doing the album in the mastering house, and I might have been spending two or three days working back and comparing, taking it out and listening on different systems in different spaces. There's a definite time element involved in the process, which needs to be considered as well. It's not about the gear; the gear helps, but you need to understand what you're trying to manipulate before you can effectively use any tool to change it.

One thing I say to students I work with is, "It's very hard to criticize mixes/masters unless you understand the entire process. You need to know what's actually transpired in the whole creative process. You don't know how many hours engineering has been spent getting the mix to the state we receive it. Unless you know the whole path from where the first sound is captured until the end of it, you can't really critique that in an effective way to say whether or not something can/should have be improved." It's very difficult, because there are so many variables. But you can clearly make comment on aspects you'd like to improve.

When I started working with people who, from a producer's point of view, wanted to work with me, I ended up finalizing their album. From a budget point of view, we got a better outcome because we weren't spending as much financially. At that point in time, going to the Townhouse, you would spend a minimum £1,200 on mastering, whereas that would effectively give us another week in the studio where we were. Overall, we achieved a better outcome from it, getting in with a couple of labels locally, doing basic parts production work and outputting for people. We also acquired Red Roaster software [what became Sequoia DAW] and the first Yamaha CDR machines. We were making test pressings/promos for labels and things prior to other things happening. It just builds, and you achieve more clients. I don't think I've ever had a bit of work that's not been word of mouth because of something else I've done. I know this is the same for many in the business.

Q: You have something in and you're asked to master it; what do you do at that point? How do you begin the process?

Prior to that point I would have suggested that some mix critique might be in order if I'm not familiar with the sourcing of it/who mixed it. I'll have a listen and see if there's anything that's correctible within the time frame of the project—basic things like excessive sibilance on vocal, consistency of or use of pan, as I mentioned earlier. Metalwork elements lacking control—aspects I'd have to work hard to correct during a master. A little bit of an overview. Really, people working on the mix of the album are too close to it to have the objectivity you get on first listen. This, in my opinion, is one of the main things you're paying the mastering engineer for—an objective overview.

Q: You'll do this before you accept the contract?

Yeah. Normally, I listen to stuff before. If I listen to it and go, "The mixes aren't there," you find a way to word it nicely if there's no way of improving them, if the source is bad. Then, if there's no way of improving, I don't want to work on it. It's not in my interest or the client's to spend hours trying to correct something to achieve a mediocre outcome. I don't want that. There are other people who can get involved in that aspect, to go back to the start of the project. The mixes need to be there in a positive way. That doesn't mean there's stuff that you don't need to correct, but it's in proportion. If they're all over the shop, is it worth the amount of time and energy? Probably not ever. There are other things you can be getting on with.

As soon as I've got the actual audio we're going to be dealing with and we know what they need from an output point of view, generally it's just levelling it and having a quick listen in overview. I don't like to listen too much—I know that could sound counterintuitive, but I think the more objective you are the less you've listened, in a way. I won't listen all the way through tracks. I'm taking fairly short samples, verses/chorus, different aspects of the dynamics and tracks to get an overview. I think an overview is really important. The context of the album should be an album. It's not a bunch of tracks thrown together. It's meant to have some continuity. Otherwise, why are we calling it an album?

Q: When you listen to the loud and quiet part, are you listening to things specifically, or is it just getting a general sense?

It's just getting a feel for the overall vibe, dynamics, and tonal aspects of it. If there's distortion, that's going to become more present in the details. Equally, noise is only a problem if it turns on and off. If the noise is there, is it is there intentionally? It's very rare on a modern piece of material to need noise reduction at any point in mastering. Even if it were needed, I'd probably use dynamic expansion; it wouldn't be noise reduction as such. It would be easier to use a dynamic expander. Say in a quiet section it sounds like a little hiss becomes present that is distracting from everything else. It's really only that section—we'd want to curtail those frequencies. So I wouldn't want to apply a broadband noise reduction that's going to take it away. Even though Cedar noise reduction is the bee's knees, it still affects the transient details. Amazing for really noisy restoration work, but in modern music you'll be looking to use some form of banding expander to reduce at that point. The MD3 or 4 are good for that sort of thing on the TC System 6000.

Q: Do you tend to have a basic chain that you like to work with?

I use a mastering console, a Crookwood, for analog insert, and TC System 6000 for digital stuff, and AES RME as digital routing matrix, as well as UAD from a DAW plugin processing perspective. Analog MS matrix and trimmers, Manley and EMI compression and EQ. Benchmark, Lynx conversion. It's all completely flexible in routing. It's just whatever's appropriate. You listen to it and go; I think to have a fixed chain of processing is deadly. If you get onto the mindset of fixed chain, you're not really being objective about the audio. There are things you can apply in order that generally make things sound better—corrective first and so on. I'm not really into commenting or criticizing others' approach to doing something, though to be objective, every piece of music is unique. To say that "This particular chain of processing will always work," is that you've failed before you started, in my mind. Obviously depends on what you define as "work."

Q: Do you rely on a metering system at all?

I like the RME meters of their hardware. I've tried lots of different metering systems. Really I just want true peak metering and varied RMS scaling. I've pretty much always used RME and some Lynx across AES/EBU. I've become really used to how reactive those meters are. I really like them. They're accurate. They're also true from a point of view you're metering it from the stream on the card, not via some other DAW software on the computer. Basically, it's not got anything else in between. You see what's coming in/out correctly. If I'm checking things going down the AES streams, I know what they are, where they are. I've tried quite a few different things. I like the Radar stuff by TC. I do use that. It's interesting to see what's going on with the audio over time. Nothing I do would request that for outcome to be at a certain loudness unit in general, whereas output legislation/ingest within the TV or film world will require people to engage in that. Spotify and other media ingest require levelling of RMS more so, and I do think it is an interesting metering prospect for the future. That whole thing of time code related, so we're thinking of average over time is an interesting point [R-128]. Equally, the butterfly metering by TC. I don't use, it but can see the use in some fields. I kind of like the concept of it. It's a useful way for you to visualize things. I don't use analog metering from a VU point of view but have them for gain reduction on hardware. Overall, for me it's just about understanding where the true peak and RMS are in stereo. Past that, it's all about the sound.

Q: Do you peg to a particular meter value? Do you ever get a mix in and, for whatever reason, the balance somehow confounds your hearing, and you use visual information to adjust what you're hearing?

No; I think relying on the visual in audio work is deadly—supportive, yes. For me it's basically turn off the "telly" [monitors], make everything grey, and then you're in a much better place to judge what's going on with the sound because you're actually using your ears! I judge level on level! As in how loud it is. That's because I'm used to the loudness of my room. Now, that comes from referencing lots of music to gauge that. So equally, I'll be using those references when working somewhere to gauge loudness. You need to be familiar with what works from a listening point of view. My listening level is between 85 and 83 dB, which I can gauge without issue in critical spaces. Once you've a ref level for this off the console, you can keep referring back to that to maintain equal loudness in comparatives—key to achieving positive results.

For metering, you'll be using from a loudness point of view. "Is it proportionally okay? Is there something about it that's a bit odd? Is it hyped in a way where the RMS seems really high relative?" But again it can be all down to the music. You can have something acoustic where you've simple vocal, which can be very open, and there's are lots of sustained pauses between the transients, it can sound very loud with only a fairly low RMS reading, relatively. The metering is not helpful at that point because what we should be listening for is an actual comparative volume to other music.

I tend to do a lot of comparative listening to different music—referencing, not to match sound but to get an idea of what I think is appropriate relative to other programme material. There are things that I do from a restoration projects point of view that are on average 6–7 dB lower than they would be on a more commercial type tune. Because the nature of the product is that we're trying to take something that was already good, clean it up, and try to pull it back to where it was then and deliver it as something people want to reinvest in. We're not "digitally remastering" it to make it "modern," like a lot of people did in the last few decades. We're resorting the audio back to its original glory, with a hint of modern digital dynamic control.

Q: Can you talk a little bit about remastering and how it relates to mastering in general?

If you're entering a restoration, especially from a point of view that the source material is poor, I've found that the best tools are the Cedar stuff, without question. The de-clicks and de-buzz/noise bits are brilliant. They all reduce transients, though. There's an awareness at that point that you have to replace that in a way to try to get back to where you were. That can either be some form of upwards expansion or normally a combination with dynamic equalization/expansion. Subtly trying to pull that back—so you're rebalancing the overall tonal shape of the audio. At which point there's a judgement call, especially if you're working on an LP, of being sympathetic to the original that was effectively mastered audio when it went onto vinyl originally. In context, are there things about that we could improve without losing any of that original work? Examples of that might be sibilance, where the nature of the cut, at the time they would be aware of trying to reduce that to a level. Obviously, we're much more able now with digital systems to control that in an effective way. That can be helpful from the overall outcome of the audio. Again, things like within the lower-mid, the bass, a little bit more control on that, you can end up with a solid low end—not in a way of trying to make it more compressed but a way to make it more even in tonal balance when it's then played back for people in a modern context, i.e., on a good hi-fi, it will sound better. The music comes across better. You've got to bear in mind that it's already being worked on, so we need to be clear why are we applying things. The majority of the processes, I would say, are to try to bring it back to where it was rather than try to make it something different—that's a good way of thinking about it.

Q: Do you find yourself working differently for different formats in terms of balancing or any sort of work?

Level is always a consideration. I think that digital limiting is something that is there in the end for loudness. To me, mastering is in the analog domain, which means our focus is with the RMS, not the peak. If you've the tonal and dynamic balance right, you'll have a good natural RMS to peak ratio. That's what I'm concerned about, but when we've got that back into the DAW. We are thinking about delivery; if we are delivering it for CD production or now more so Mastered for iTunes. Generally, we're focused and considered about where the peak level is relative to 0 dBfs. You're avoiding any clip of the encoding process to Apple lossy delivery or possible DA conversion error on CD playback.

Q: What about "FDRHD"?

Full Dynamic Range and High Definition is my effort to help the world achieve more dynamics in delivery—to hear what we, as mastering engineers, are privileged to hear. The principle being for me is when I'm working on audio in the analog domain at the end of the transfer path before capture back to digital, we've massive headroom to work from. When printing the audio back into the DAW via AD, we're capturing it back at max peak before clip. The restriction exists when you're converting it to digital. There's plenty of headroom still in the analog realm. That converted print, to me, is the best it can be in digital format. That's what I call the FDRHD, because it is basically at the highest resolution it can be in terms of the capture relative to best sound and playback delivery resolution and dynamic range. There's nothing been applied to it to reduce it because of the delivery format or a want for perceived loudness. This would be a minimum of 24 bit, 88.2 kHz or higher going up to DSD, which I'm currently experimenting with.

At 16 bit delivery, we're going to be applying dither to the above for good reasons; otherwise we'll end up with quantization distortion. If we're reducing it in terms of sample rate, which again can affect it depending on how we're converting it. With regards to the whole thing of loudness, I've always been a believer in the volume control. The limiting bit is because of the "louder sounds better" scenario. That's not to dismiss it from the point of view of that's what the delivery of the marketplace is. We should be doing that as required until the marketplace changes. That doesn't mean that there can't be multiple ways of delivering that, either. One of the things about Mastered for iTunes that is appealing is the potential for dynamics. Obviously, what's not appealing is its still rubbish lossy quality. I'm not saying that I'm anti-limiter. If we take something like, very aggressive rock, part of the point of why digital limiters are very useful for me is the controlling of those transients and making them even in terms of kicks and snares, especially to get a very smooth, even edge to the top end—that clear definition between the kick and the snare and their harmonic detail between them. From that point of view, you have to apply quite a lot of limiting across the chain to get it to a point where you get that roundness. Otherwise, differing parts are moving forward in the mix relative to others.

The TC System 6000, BL2 does it really well because of the soft clip. It cuts it and pushes the transient shape. But what I really like is the work I've done in the analog domain. When I've captured that back in, that's the best it is, and that to me is the full dynamic range and high definition. I'm not applying anything to it, and I don't have to. Everything I apply to it from that point is normally a compromise because of delivery format. If you take the best capture which is more dynamic at equal loudness with a version more compressed/limited/rate reduced, I think anybody on a good hi-fi system listening would prefer the one that's more dynamic. Transients are good!

Though, in saying that, the argument about loudness is about how people digest music. It would seem a lot of young people [are] listening to music on ear buds, iPad, or laptop speakers. For us, being an older generation, music used to be a more communal thing where you'd listen to a record with people on their hi-fi. That communal thing still happens, but it tends to happen with rubbish playback devices. It's literally ineffectively laptop speakers, Bluetooth mono sound blocks, or bars. Even when we were listening to albums on portable cassette tape machine, at least it was played back on a speaker that was about 3 to 4 inches, so you actually got some bass. Even if it was mono and rubbish, it still had bass. There's a lot where I think the young generation currently find it difficult to ascertain what quality is, because they never heard it. If you've never seen green in the sunshine, you cannot describe what green is.

Q: The younger generation, when they learn about sound quality, they're "educated" by their playback devices. If you have this generation that doesn't recognize the difference in sound quality, then how do you get across to them?

I don't think you need to. I think it comes from exploration. I think that's just the nature of how things are at this point in time. You see there's a lot more people into vinyl, which means they've got to have a deck, and inherently they get a hi-fi. They tend to do that on a budget. Students go and buy old hi-fi separates for a few hundred pounds. It might have some nice old Celestion speakers, which actually sound pretty good compared to your little playback system you get from the hi-fi stores. It's the nature of people to explore. It's like being a guitarist. You've got this rubbish little practice amp and your guitar. You work out that you can't play the guitar really well because the neck's rubbish. Then you get a much better guitar, and actually it still sounds rubbish. Why does that record sound like that and this guitar still sounds rubbish? You buy a better amp . . . and so on. I think it just takes longer for that discovery to happen than it did when we were younger because of the initial interaction with music. That doesn't mean I don't think it will happen. Inherently, I think things will get better in terms of the technology for sound. You've got those speakers you can drop on the desk, and it generates bass from the surface it's on. I think innovation will catch up, to a certain extent. I don't think we need to change anything or criticize anybody. Just let people enjoy music. The more they enjoy, the more they want to discover it. What we should be doing (as always) is trying to maintain the best that we can. I think there's a natural thing within people creating art, where you want to do it to the best that you can anyway. The people involved in it are relevant to the output and would always be trying to do it really well. One of the people I work with always says, "If I'm not making an album for myself, who the hell else am I making it for?"

I'm more pessimistic about the revenue stream of how people get paid for music than I am about the quality of how people listen to it, because I think that will improve. I don't see a positive way out currently for the financial devaluation of music. Companies and people are accepting it's okay. But that's across all areas of business; people's accumulated skills are not valued in the current economic environment.

Q: Currently, this is a generation where music is free online for them. It was never an expenditure. The industry itself and the services

**within the industry will have to accommodate somehow. With dwin-
dling revenues what do you see as the future of mastering itself as
an industry?**

I think that in the same way that I suppose that there aren't the same
number of studios that there were, because of the nature of the upkeep
of them and now the technology permits people to record things themselves to a high level. Equally, producers have moved into the realms
of, "We'll go to that person's house and record. We'll make the right
sounds with the right gear, which we'll hire in." It's that change. It's
more of a cottage industry aspect to it rather than, "Oh you've got to hire
this particular space." The same thing is happening within mastering as
we speak. There are many mastering engineers who work from a place
that wouldn't be classified as a commercial space in the classic "mastering house" format. The mastering studios I used to go to as a young
man don't exist in the way they did from that perspective. I think that's
already happened, in a way.

You always need microphones if you want to record something that is
live. I think artists discover the technology and start to use it. It's a natural
realization, especially within the band community. Eventually you need
someone to help you record and mix. "I want it to be as good as those
things I'm listening to, but we're not achieving that. To do it we need to
get someone with the skills to do that." That doesn't necessarily mean
you're going to a studio. You're employing someone in to add that skillset
on. I think at some point those people doing it find it still doesn't sound
how they want it to sound. They then go, "What I actually need is a mastering engineer." It's a natural progression.

Inherently, record labels, producing a lot of stuff, need people who can
just deliver the outcomes they want, how they want it, at quality, and completely prepared for delivery. A commercial job, 60% of it is parts production, not the actual mastering in terms of changing the sound of the source
material. Basically, I'm spending twice the time producing all the different
versions required in relation to the sonics.

I don't feel pessimistic in any way about it. I suppose when I came into
it, it was already changing massively. I was part of that first generation
of people, almost knocking the nail into the coffin on tape-based studios.
I didn't consciously think about it at the time, but if you look back on it
and think about what happened 25 years ago, that's what we were doing.
We bought into the digital realm of stuff in a way that you could do things
that you couldn't do otherwise. There were things compromised because
of that, but there's that natural progression of technology that's constantly
in the music industry. You have to stay on top of that. If you're not, you'll
just go bust. Even then when there was a lot more money slushing around,
if you were in it for the money you'd not be making much! You have to do
it because you love the music—if you love money, become a banker. The
industry is always in flux.

Q: It's almost a state of constant transformation.

Yeah. If you look at the TV industry at the moment, there's all these independent regional companies that coming up out of the new licencing even though we got rid of them all 20 years ago. It all comes about because you can now buy [an] HD camera for a few hundred pounds that's as good as something you would pay 20 grand for not that long ago. It's the same thing that happened to the music industry 25 years ago that is now happening in the film and TV industry. Work that classically had to be completed in a commercial post studio can now be done on a laptop. One of the great growth areas for the cottage industry engineers is being able to work on that post stuff.

The artist revenue thing is still something I feel pessimistic about, because how do you actually make money out of the music you made? You should be getting remuneration for what you've created. I'm not anti-streaming; I'm anti-single track! You used to make albums because they are a musical journey, the same way we used to have classical music: you'd have a set of movements because they're meant to be enjoyed as one. An album is meant to be enjoyed as an album. That's where it went wrong as the major labels clamoured to get on iTunes. They forgot how they made money—selling physical albums—why stop that just because you can download it! Simple really.

I read an interesting comment the other day. Basically you'd never go to a party nowadays and take someone's vinyl off the turntable and change it, but you'd quite happily go over to someone's Spotify or Apple music on their iPad and flick the track. That's quite interesting, because I can see that quite clearly in the younger generation. There's something more precious about how you deliver the medium of the music, as well. The more we can get back to album type/tangible delivery, the better the revenue thing will get, in my opinion. Basically, saying no to single-track download, the more we'd get back in the long run. I know it's not as simple as that, but that's an ethos. Listening to music is relaxation, we should get back to listening to the artists imagination of music, their musical journeys—albums are key.

Matt Colton

Matt Colton was awarded the Music Producers Guild Mastering Engineer of the Year award in 2013. A mastering engineer since 1997, Matt joined Alchemy as a director and engineer in 2012 having previously worked at AIR Studios, Alchemy Soho, Optimum, and Porky's Mastering. Matt has worked with many of the world's biggest and greatest artists, including Coldplay, Muse, James Blake, George Michael, Peter Gabriel, Manic Street Preachers, Leftfield, and Flume.

Q: How did you become a mastering engineer?

I started in radio in my native South West under the equivalent of the then-Youth Training Scheme. The station thought it was a good idea to get an intern to write the scripts to the adverts. We soon realized that I was not very good at it and that I was more interested in the gear: the cart machines, the Revox reel-to-reels, and the like. Soon enough I was working on that side of things and getting quite good at all that; then, towards the end of my time there, I was doing a Saturday show, and there was a weekend presenter, Mike Marsh, I knocked about with and went down the pub with. We went out one Sunday for a beer after his show, and he got a phone call on his brick of a phone (it was 1995), and he said, "So they want to do it next week. Can you call them back and say I can't until two weeks later as I'm holiday?" So I asked Mike what he did, and [he] explained what he did at The Exchange and that he was asking The Exchange to let U2's management know that he couldn't master their record for two weeks later than they scheduled! Mike later said that I ought to come up to The Exchange and hang out on the sofa at the back of the studio to see what went on. I'd already been playing in bands, sequencing stuff, and recording a bit, but I'd never heard of mastering. In those days before the internet, it was difficult to find out about things with the ease that we do today. To make a record, you learned on the job, and there was very little way of knowing what mastering was unless you were introduced to it.

As I spent that day in Mike's studio, I realized that I needed to do this. So I asked him how to get into mastering. Mike said he'd no idea really, but suggested I buy *Music Week* and see if there were any ads in the back. Off I went the next week to get the latest copy, and in the back of the magazine was an advert for a trainee at Porky's Mastering in London. I started by just reviewing masters and checking for errors in real time and making tea, but after a while I was doing small compilation albums and the like. Not full mastering as such, but levelling and making sure the PQ data was all correct. Eventually I rose to start fully mastering.

Q: How did you learn to master? Did you have teachers?

Paul Solomons was an engineer there at Porky's Mastering, and he taught me how to cut vinyl and made me think about mastering in a much broader way. From there, everyone I've worked with or shared a room with has taught me lots along the way. The biggest influence would be Ray [Staff] who, once he decided I wasn't an idiot, gave me lots of time, and I bet there are lots of my peers that would say the same of Ray. They, like me, owe him a lot.

Q: What is mastering?

This is a tricky one. The way I try explain this is that mastering is more about the "Why" than the "How." I put it like this: when I hire a plumber, I don't want them to just have a collection of spanners and what not, I want them to know the reason to use them. I want their expertise to know which tool is right for what job and "why" they need to use it. I can own the same tools but would not know what to do. Too many people consider themselves to be mastering just by using the tools we might use. It's the "why" you go to those tools that's important rather than the "how."

Q: What are the things you typically listen for. What aesthetic areas might you address in a mix, regardless of genre?

As you say, I can usually tell what needs doing within a minute or so of listening to the music. But the more I contemplate, the worse my judgement can become. I second-guess myself and become less sure. For a mastering engineer, it is that ability to listen and judge quickly. It all boils down to low, mid, high, dynamics, and timbre or tone, really. Those are what we adjust, and not by much a lot of the time.

Q: Do you feel the so-called "loudness wars" is over, as many are now suggesting?

I think it is the way people are listening now that will eventually change the loudness wars. For example, a decade ago, everyone had to have their iPods with thousands of songs on, and it was important to carry the whole library about. Without any level checking, you were on shuffle and turning up and down the volume all the time. Now that iTunes and Spotify, and it seems YouTube too, have level matching, it no longer matters so much whether you're all as loud as possible. R-128 is starting [to] educate people. Vinyl, too. That's always been dynamic, and it's a little ironic that we need to get things loud for vinyl to get a good signal-to-noise ratio, and yet it frequently has more dynamics than digital releases. As people are now engaging with a format that they have to get up and place a needle on the vinyl and they have to turn to side B, people are listening in a different way again.

There was one client I had in a year ago with Ableton on his laptop. He said to me, "I've mastered it on the bus here." I played it through our ATCs, and I felt the track was a little swamped, and then looked at the bus. It had something like 10 or more processors on, including three multiband exciters. No wonder the dynamics were squashed. That's the thing these days. Whether it's because of the formats being more accurate, we are used to listening and appreciating really fast transients. In the old days of cassette and vinyl, things were maybe a bit smoother, but now if any of those fast transients get slewed, then the track becomes lifeless and swamped. So I asked him to take off all the plugins, and I listened again. Suddenly the music came to life and had space and dynamics. We then placed an EQ, a compressor and a limiter and did just what was needed for the track. It sounded pretty amazing in the end. There are too many people who have the "how" but not the "why."

Q: Where do you see mastering heading in the next 10 years, both technically and business-wise?

I've been mastering 20 years now, and around 10 years ago I felt that the industry was in decline, and I began to discuss with my girlfriend what I could retrain in. I wondered whether in the future I would have to earn a living doing something else. However, I dug in, and I've worked at places like Alchemy and AIR mastering, and I'm now busier than I've ever been. I'm doing great stuff with exciting artists who don't seem to be going away anytime soon. I'm hopeful.

Nick Cooke

Nick Cooke is an independent mastering engineer and professional musician. He received his BA Hons. degree in music production from the Leeds College of Music and started mastering in the production music library industry in 2008. Nick's mastering career began at De Wolfe Music, which was founded in 1909 and is the longest running independent music publisher in the world. He quickly built his skillset working on a vast and varied catalogue and specializing in remastering projects from tapes dating back to the 1950s, as well as newly produced material. In 2010, Nick moved to Sony/ATV's award winning Extreme Music. During his time at Extreme, Nick has had the pleasure of working with many leading composers and producers such as Hans Zimmer, Sir George Martin, Hugh Padgham, Ramin Djawadi, Eddie Kramer, Timbaland, Xzibit, Michael Giacchino, Lorne Balfe, and many more.

Expanding out from the Production Music Library business, Nick has been building his independent mastering portfolio and has mastered material from many independent artists, many of which have gone on to win or receive nominations for various British music awards. In 2018, Nick moved back to his native South West of England to open his own mastering facility in Bristol. In addition to mastering, Nick has worked as a musician for over 15 years and currently tours and records regularly as a professional accordionist for the multi-award-winning Kate Rusby.

Q: How did you become a mastering engineer?

By fluke and luck, because I didn't apply to be a mastering engineer, and to be honest, I didn't know much about it at the time. At university we covered a bit, and I got the opportunity to meet and have a conversation with Bob Katz as well, but apart from that I didn't really know too much about it. However, I saw a job opening as a music editor for a company, a production music library called "De Wolfe Music" in London. It was a great interview; we chatted about Glastonbury Music Festival, so, most of my half-hour/40-minute interview was just laughing about Glastonbury Festival, and then a day later I got offered the job. I turned up and got put in this room with a pair of speakers, a couple of bits of outboard, a tape machine, and SADiE 4, that I'd never heard of before, but that was it. I was asked to compile a bunch of tracks into an album and do any processing I thought fitted, really, and that was my introduction to mastering. So then I went home after that shell-shocking day, going, "I have no idea what I'm doing!" and then dug out any books and anything I could find on the internet or whatever, anything I could think of to learn a bit more about mastering. With that, it sort of just developed; I've never had any formal apprenticeship or anything—I just sort of dove into it by accident.

Q: So then, how did you actually develop your technique? How did you figure out where you were supposed to go with the material?

With that stuff, as well as working from freshly composed, freshly mixed material, a big part of this job, and I'm really fortunate for this now, was remastering off of tapes. De Wolfe go back to 1909, which is when they started, so pre-recording, they started in the sheet music production and publishing side of things, so eventually they were doing music for silent films and such, for small ensembles or pianists to play in cinemas, in the silent era, which I find amazing. And then obviously that progressed into recordings, onto tape and then onto vinyl, CD, and now online. So I think the oldest tapes I was working on was probably the late '40s, which I find incredible. So I learned about the whole "baking" thing, and my job was to get across as best as possible what was on these tapes without screwing it up. Hence, I guess the way I developed was just listening and listening and listening, so much listening. I would spend my lunch times listening to stuff I wasn't supposed to be listening to—it was just that, just getting an idea and a feel for things. Mastering is more of a feel thing than the technical side of things for me. You have to learn techniques for all the transferring, so I learned how to balance tape machines from others around at De Wolfe, along with trial and error, as well as all the de-noising stuff and all that, but most of the time it's just listening.

Q: Yeah, a common theme we're picking up on is that the most important gear in the room is your ears.

Yeah, your ears, your room, and your familiarity with your surroundings: I'd say that's the most important thing. Even to this day, I still go back to my folks, to my parents, to my dad's old hi-fi that I grew up listening to. Every now and then I'll stick something on that I've been working on or something new, and just go, "Oh, what does it sound like on this?" Just because it's so familiar to me. People talk about how they listen in their car or whatever, but just use anything. You automatically know how things should sound with it. I love that; I love the way the whole human mind works with that, that we're built on familiarity and experiences—that's how I see it.

Q: If you had to, how would you define mastering?

It's kind of tricky, and I guess it's developed over time. I guess how I see it today, I still see it basically how it traditionally is, as in getting a recording from one format to another and making it sound as accurate and correct for the artist, producer, engineer, and the performers as possible. So back in the day it was fitting whatever was on tape, or whoever was in the room, because they would cut straight onto disc sometimes, whereas now I deal mainly with online stuff in my day-to-day work here at Extreme music, because we don't produce CDs anymore. It's all online at broadcast standard files (48 k 24 bit), but also it has to fit. You just have to imagine where it's going to get played: I've heard a track I've worked on being played in the Olympic Stadium, and I thought what the hell, so I've got to cover for that and as well as on someone's phone speaker, or radio, or someone's computer toy or anything. I guess these days it's maybe a little bit more—I wouldn't say it's more challenging, it's just different; there's more varieties on how people are going to listen to it. So I think to answer the question, it's kind of that link between creativity and technical, so it's the last stage of the creative process but making it fit in the technical world. Although mix engineers, recording engineers, even producers and a musician will think about the technicalities as well, but the mastering stage is really the final place, or if you're cutting to vinyl, the cutting engineer is the final sort of creative technical stop before production and release.

Q: So basically it still is deploying a kind of creativity to ensure optimal transfer to whichever format the recording is going to be distributed on.

Yes, exactly. Whilst also obtaining or maintaining the optimal sonic quality and the style and feel of the material as well.

Q: Are there any mastering sessions in particular that you remember as having been instructive to you?

I've been very fortunate working at Extreme Music, who really allowed me and supported me to develop further as a mastering engineer. I've been working on all sorts of different styles of music being produced from quite a wide, varied collection of producers, engineers, and composers, ranging from your A-List Hollywood composer and legendary producer/engineers to people who are just starting out. Either way, my approach is generally the same. Mastering to me is a communication point, as well; it isn't just an engineering and listening vocation—you have to communicate with people and understand what people want. And I think for me, one of the first times that happened where I was kind of going, "Oh, am I doing the right thing?" was when we did a series of albums here with Boris Blank, who is one half of the group Yello, so he's a sampling godfather, and I was shit-scared anyway of working for him because he's got such a reputation of being so spot on with his mixes. I guess that was one of the first truly great mixes I had, so to hear a decent mix where they were so certain of their sound, that was really great, because it made me aware of what mastering was about more than when I've had to fix stuff with mastering. I was there literally to enhance stuff, and then as an album point of view, to balance the album, as well. It was really a kind of gentle mastering process, which I got so much more out of than having to carve into tracks, not that I tend to do that anyway, but it was the first time I really had to do very little, yet understood what mastering could really bring.

Q: Any other sessions?

I mean, tape sessions—I'm still amazed by tape, and I miss it, I don't get to work with tape anymore—it's got that physical quality to it, and it's just got a bit more history to it. I guess with tape is just easier to understand, because it was a bit more physical than an audio file, like you can literally see the needles moving around, and you can almost see the audio turning into electricity and back out again, and you can hear, it's almost like capturing something. I find that amazing, I still do; I just love the quality of it. Working with tape was an eye-opener for me. Then again, with more recent stuff, it's just seeing people's reactions. We had a composer in who had travelled down from up north to do a showcase in here, and I hadn't ever met him or spoken to him. He was just so grateful for the work we do here, and that's awesome. That's what we do really, so hearing little stories from people being grateful of the work we do, and going, "Yeah, that's how we wanted it to sound," that's when I've gone, "Ok, we are doing good," because a lot of the time you're just shooting in the dark, but to get the reassurance is nice.

Q: What are the things you might typically listen for, regardless of genre?

I guess it straightaway depends on whether you'll be dealing with a single track or an album. If you're dealing with an album, especially in the library world—I know it happens in the commercial world as well—I'm dealing with four, five, or more different producers, mixers, engineers, across one album. So getting an idea of how an album should sound, you need to listen to everything to kind of work out in your head what the hell is going on, or you speak to the producer or the artist. Actually, with albums here we would have several producers, and then one overall producer on the project to say, "Okay, out of all of the tracks you've got here, what's the closest thing, you would say?" That's for an album. Then thinking more specifically, I would just listen to a track and listen out for clarity, seeing what I can hear—if there's anything poking out too much, I guess, in terms of frequencies but also in terms of dynamics, like certain hits and that sort of thing. Is there any kind of hiss you need to think about, or noise, or noise floor, or what's the key thing to the performance, as well. So if you're listening to, say, a jazz trio, then you're not going to be listening out for a vocal line necessarily, or if you're listening to a pop track, you're listening for not just the vocal line, but what's the hookiest bit—is that the riff or is that the vocal line?— and stuff like that. You're kind of thinking about the composition and the music side of things and where the focus could be. I always say I'm a musician first; I'm still a touring musician, all over the shop, so I would see it from a musical point of view first, before a technical point of view. I think that really does apply to mastering, because at the end of the day, that's what a master should be, you need to get the music across and the musical point across. It's all about feel rather than a technical thing, for me, anyway. If someone's sat with me and watching what I do, they may say, "Oh no, that's technical, what you on about?" But I see it as a feel or hear it as a feeling type of thing.

Q: Do you have a preferred sort of chain that you use when you master?

I guess things have stayed in place. I've got a hybrid setup, so I've got various analog outboard bits of gear going through an analog master transfer console, a Maselec unit here, as well as digital outboard and plugins, as well. That's important: I've got digital outboard processors as well as plugins—for me, digital outboard can sound better, and generally I prefer them to plugins, although plugins really are getting better all of the time.

Q: Can you take us through the spine of the chain?

Yeah; going from the computer I would go out to the analog chain, I would have my Analogue Tube AT101, which is a Fairchild copy compressor. It's a tone box—that's how I see compressors anyway, they're tone boxes—so I do that, and then because that just kind of opens up the sound and does something to the sound straightaway, I would listen to that, and then I've got a Maselec EQ here, so that's a really high precision EQ, so I might use that afterwards, depending on what's going on. Before that compressor, thinking about it, I might use a Weiss EQ here, and I might do some precision stuff with that to tidy up the material. So if you've got a crap load of bass going on or certain bass frequencies that are pushing the compressor around too much, then I might reduce that first to allow the compressor to do its job better. Thus, rather than it compressing, it's just the sound of all the tubes and stuff. There can be certain frequencies that can screw up the tubes, either setting them off, distorting the signal, or can make them sound a bit weird. It's just a case of getting those to sound as best as possible, and then I might have another compressor after that, just more for the gain stage. I feel weird calling them compressors, because I don't really use them to compress that often.

Q: Can you elaborate on that?

I see compressors—especially the two main compressors I use here are these tube ones, so the Fairchild copy, the Analogue Tube AT101, and then I've got a Manley Vari-Mu; they each have a very distinctive sound, it's a good sound, they're two very different qualities—they're tone boxes, and from experience from years of using them, I've got to know the way they affect material, as well. That doesn't mean to say I stick it in and it sounds exactly how I think it's going to sound. I might have to take it out, do some EQ beforehand, just to make the tone of it work well. I kind of always see them like another type of EQ in a way, but it's also to do with gain staging as well, because you go through tubes, transformers, and circuitry, and it will just change the sound even if it looks like it's doing nothing—it's always doing something, it always is, so that's kind of how I see it. So after saying all of that, there are of course times when I haven't used them, as they change the sound too much for the material.

Q: What about for conversion and monitors?

So, monitors, I'm on PMC's here, PMC IB1S, run with the now-classic Bryston amp combo. Before, that was one of the first things we changed when I joined this company, they were on Dynaudio BM15's, and it just sounded boxy as hell to me. So the first thing we did, or I did, was to get some acoustic treatment in the room. We've since moved facilities, but in the old room I got some acoustic treatment and got these speakers, and it just completely changed the consistency. I wouldn't necessarily say the quality of mastering, because they did do great work before, and I don't always believe you need the most amazing equipment to master. I think it was someone like Mandy Parnell, who I presume you're going to inter-view, she said, "You can master with a Mackie desk," or she might be able to, but as I said earlier, it's due to familiarity with your room and your conditions and having something a bit consistent, and so by putting acous-tics into a room, you're aiding with that. So I just made the room more consistent so I was able to do more consistent mastering, because that's really important for a mastering engineer, which is to have consistency, especially when you're working on an album or several albums a year, or a week, or whatever. Then going back to converters, I'm just on your bog standard Avid I/O the latest ones, which admittedly, that is what I would like to change—I'm working on changing that at the moment, but it's an ever-going project to build this stuff up. I sort of know what the converters do to the material, and with my transfer console, I'm able to hear before and after the converter, so I can hear what's going on and maybe address what's going on.

Q: And your console, that's a Maselec?

Yeah, I've got the Maselec MTC-1X. They're incredible! Leif, who designs and makes all this, he's just such a nice guy to deal with. I think I have one of the first MTC-1Xs because I wanted the whole gain adjust knob for A/B-ing, so I can have my original track come in on another channel or source input and match the gain, which is just crucial. I was surprised he hadn't put that in in the first place when he's got that on all his other transfer consoles, but he's done that now with this unit, and I think they're great.

Q: Regarding broader equalization concerns or broader approaches: if you can talk about the way you think about equalization in mastering, as opposed to even when you're performing or recording, does anything sort of, or is there anything there?

EQ is so important. You can go overboard with EQ really quickly, I believe. I often think if I'm putting on a ton of EQ, I can stick my Weiss on, do a bit on there, and then stick a bit more on my Avalon EQ here, and then do a bit on the plugins, and it can just start to sound a little bit disjointed. So I usually strip it all out, and then start again with broader things, or maybe just one EQ. When I started mastering, I think it was the best thing for me, I only had two bits of gear to use—I had a Weiss EQ and a Weiss Dynamics compressor unit as well—and then SADiE, and I think that SADiE only had their master limiter on it, it didn't really have the EQ or anything, so I had to learn how to get whatever sound I could think of out of just two units. I think limiting yourself is tremendous, because if people start carving into a track, you lose nuances, the feel of the room, whatever; it just starts to sound weird sometimes, and if you find you're doing that, maybe you need to go back to the mixing stage or whatever. I think mastering should be gentle.

Q: How about stereo image? Do you address through mid/side work or anything like that?

I used to. Actually, the Weiss units here do MS, and I've got a TC System 6000, which does MS, and you can do multiband MS stuff, and also the transfer console here has an MS circuit built into it. I've either got an EQ or a compressor that's able to go through the MS circuit, so to begin with, like with any new toy, I used to play around with that a lot. However, I've found I'm doing less and less now, as it can quickly sound disjointed and wrong, so I tend to not do it. I find it's amazing what you can do with EQ to sort of hint at what you may be trying to achieve with MS processing. I do use the elliptical filter on the console to mono the subs and bass sometimes—I think it can be a really useful tool, especially when it comes to electronic music, if people are going crazy with their panning, but I'm quite fortunate with that, as a lot of people I work with don't. Saying all of that, if there's maybe some other crazy situation, like sometimes a piano is something some people can massively pan, and then it can sound wrong, so I might use a bit of EQ in MS to reduce some mids in the sides to help the focus in the centre a bit more, but it's really rare, to be honest; I try not to use it. The stereo width control I've got in this Maselec is incredible, but it's not an MS circuit; it's slightly different, I believe. I'm not amazingly technical, but I think it's more of a circuit that you used to get on the disc cutting consoles, as opposed to the MS thing. I think it's a different sort of equation or something, but it sounds a bit more forgiving if you check it in the mono, and it always seems to keep stuff mono, which is good.

Q: Do you think the so-called "loudness wars" are over?

Potentially in some worlds. The medium I work in, TV and radio, I think there are still some issues there. I know in broadcast scenarios they are putting caps on volume level for adverts and stuff like that, but still people are competing like crazy. I even notice it; for example, we had the Music Library Awards the other month and heard how some tracks were just ridiculous compared to other ones. So I think it's still there, but it is going, and I guess it depends on how people listen to music. If people are into vinyl and stuff like that, then they can tend to be against heavy limited music. However, now with things like what iTunes is doing with its Mastered for iTunes standard, that's obviously helping people think about volume levels again. It's nothing new; people have been saying for years about how massively slammed tracks and clipped tracks make bad mp3s or bad down-sampled quality files, so I think it is improving, definitely, because we're listening to formats where it's less forgiving to have a slammed master. But then again, I do get asked to make it as loud as possible, too.

Q: Can you talk a little bit about the differences in mindset on how you approach a remaster versus a master, per se?

I guess with a remaster—especially if you're working off of a master tape as opposed to mixes, which is what I did a lot of, so it wasn't necessarily the mix master, it was the master master—it's already sort of gone through a stage of mastering. But my job there is to get it into another format, and so with that, it's to get it across as cleanly as possible, just trying to make it sound how it sounds in the room, but in another format. That's what a lot of the remastering I did was; but saying that, a lot of the time, I had the master tape in front of me, which might have been sat in the storage rack for ages and needed to go through the process of baking, because all of the oxide had stuck together. You get a lot of that at the moment I think there was a batch of tape in the late '70s or early '80s that just turned out to be utter crap. I think the guys were saying that it was the best stuff you could get, but then it all went to pot. So I'd then be comparing it to a vinyl release, if there was a vinyl release of it, and then trying to get it as close to what the vinyl sounded like; it was just a kind of balance. Whereas with a newly mastered, newly composed track in this day and age, where you're not remastering, I guess you can have a bit more scope. It depends on the project, because you get some engineers and some producers who want it to sound exactly how the mix is but at a more commercial level, or consistent across an album, but then you've got other people who go, "Can you just make it sound punchy as hell?" I mean it depends on what people want; we're a service, at the end of the day.

Q: So it depends on the brief, really?

Absolutely, yeah.

Q: And with remastering, it sounds like what you're saying, it's sort of more closely aligned with the traditional transfer tradition within mastering, where you're trying to create an optimal transfer but not necessarily "updating" the balances or anything?

Absolutely, unless it's completely wrong, because you are dealing with an older format, and over time, the top end or the low end, or whatever, might have gone slightly, so that comes down to experience. The more things you listen to, the more you understand, and the more things you go and watch, like if you go and see an orchestra live, or a rock or pop gig, you have that experience of what things sort of should sound like; you can respectfully bring it back to what you associate it with. I mean I've done that in the past, and I've been shot down for it because I got it completely wrong, but I'd just redo it, so, yeah, it comes from experience of listening.

Q: You mentioned baking; could you just take us through what that is, and how you've come to experience it?

Yeah, it's a special humidity-controlled oven—you probably shouldn't do it in your home kitchen. I mean, I haven't personally done it—I've had tapes sent off to be baked—but it's where the oxide of the tape has come apart and the tape becomes what we call sticky, so it starts to fall apart and gets stuck on the back of the tape it's attached to, and things like that, and doesn't go through the machine particularly well. Hopefully, you notice it before you try it in the machine, because unfortunately, I've seen it happen, and it just comes apart, and that's it, so you've lost that tape and the recording. So it's a process: it's a certain temperature, fairly low, just to remove the moisture and reactivate the binder that sticks the oxide to the plastic; that's done over a period of time to just help loosen the tape enough to be able to be played. I remember people telling me, I mean we used to use FX Rentals in London, who are very well known, I think they did all of Bob Marley's back catalogue a few years back. There was a story about how you're only able to play it once after you've got it back from baking, but apparently that's a load of rubbish. I mean the tape's playable life is limited, and it is a bit more delicate than, say, a freshly recorded tape, but from what I've been told, you can play it more than once as well as possibly rebake. It's magical what they do, anyway, whatever they do.

Q: How has mastering changed since you started doing it, and where do you think it's headed?

I still consider myself pretty new to mastering, but I've been mastering for about eight years now, and every day, every track I do is completely new, and I'm still smacking my head against the desk going, "What the hell is going on?" half the time and trying to wrap my head around it. However, saying that, from speaking to more experienced engineers, that never stops. So where do I think it's going? From my personal journey, it has changed quite a lot, but I think my journey is quite backwards, because I started off doing a lot of remastering and working from tape, whereas I guess nowadays people start with their mate's files or whatever, but I learned from a very analog point of view, which I'm grateful for. I guess that's the traditional way: speaking to your Bob Ludwigs and people, that's how they started, I guess, so I'm hoping to develop and go back into doing that, that'd be great, and get into cutting records as well; I'd love to do that. I'm hoping that'll happen, because certainly the vinyl market seems to be increasing. Although I haven't seen the numbers for this year, but certainly a couple of years ago it was up 30-something percent in sales. But I think technology as a whole seems to be finding ways of making better-sounding devices for the mass market, and I guess as technology becomes cheaper and parts become cheaper. However, I think there are many people stressing how mp3s and all that are killing the audio world, but I think it's actually going to come back around because people, at the end of the day, still like listening to music.

Mastering for me has changed, because I've been able to develop and build more of a studio and pick equipment that's best for me, especially at Extreme. I've learned a lot about outboard; although I've done bits and bobs in plugins, I've never been obsessed with plugins because I like the physicality of outboard and the connection and experience you have with the music. Then there's the sound quality: once I started using analog outboard gear and I heard what they could do, I was hooked! So I think in terms of whether plugins are taking over outboard, I've not really personally heard plugins that are doing the same job as a really outstanding bit of outboard audio gear; there's something about going through circuitry and stuff that really does something to a track, although you wouldn't necessarily notice it—it just seems to open it up. So I don't really see the outboard side of mastering going away anytime soon, because there are more manufacturers and people seem to be getting excited about it again. The whole world works in a cycle, like we've had '80s music and now we're going back into Brit Pop again over here, so everything kind of cycles around. We had Motown-sounding stuff not that far back; it will keep on coming around.

I think there is still a need for mastering as well, especially when you consider the number of different formats a piece of music can be released on. It's even more crucial to get someone who understands all of that. But I guess I'm somewhat biased!

Q: If you could go back to the very beginning of your career and tell yourself one thing, what would that be?

Relax, seriously! Because the number of times I've come home from a day of working on someone's track, and gone, "What have I done, I've just ruined someone's piece of music?" or "It's just not working," or something like that, I go in the next day after having like a massive fit or something, and then gone back and looked at it again. It's amazing how the second or even the third time of doing something, if you look at it in a more simplistic view, it can just be a lot better, so I think just relax and chill out.

For me, a well-mastered record is that I never want someone to listen to a record anywhere and say, "Oh what an excellent mastering job." I'd much rather they talk about the music: that's a well-mastered record for me, if you're listening and talking about the music.

Paul Baily

Paul learned about music production using analog technology, creating pieces of music with thousands of pieces of analog tape edited together and a row of five reel-to-reel tape machines. After completing a master's degree in composition at Durham, Paul began his career at BBC Radio York (still recording and editing on analog tape in those days).

He joined EMI's Abbey Road Studios in 1986 as a classical editor and by the time he left in 2000 was in charge of classical music post-production. He worked with artists in various genres, including the Pet Shop Boys, Nigel Kennedy, The King's Singers, Sir Simon Rattle and many more. He has worked on film music scores including *Mighty Joe Young* and *Captain Corelli's Mandolin*.

Paul set up his own company in 2000, based in Thirsk, and specializes in remastering old analog recordings, mastering, and location recording (engineering and producing). He is the only third-party engineer with permission to take EMI and Decca classical master tapes off their premises.

There are approximately 2,000 CDs on various record labels which have been remastered or mastered by him. Several have won awards. In addition to being a part-time lecturer at LCoM (Leeds College of Music), Paul is also a piano teacher and an A-level Music Technology examiner for Pearson Education (EDEXCEL). He is a classically trained musician and plays violin and flute as well as piano.

Q: Did you always want to be a mastering engineer?

No! I fell into it almost by accident. Having completed a music degree at Newcastle, I went on to study for a master's degree in composition at Durham. On graduation, I still had no idea what to do and ended up spending a year looking after extracurricular music actives at a school in York. Next door to that was a local radio station, BBC Radio York. I went around there and said, "Give me a job." They said, "Okay but we're not going to pay you." I got in on that basis, but soon they could see I could do a few things, and I started earning some money. This was back in the late analog days of 1984 and 1985. They'd send me off with portable reel-to-reel machines to make vox pops in town. I'd do interviews and edit them together with a razor blade and splicing block. When I was at university it was all analog, so I was very proficient with tape editing. After doing that for a year, I typed about 200 letters to every record label and studio I could find in the UK. I got one positive response from the director of A&R in EMI Classics. He put me in touch with Abbey Road, and I talked my way in, starting as an editor on the earliest digital editors. I soon moved into remastering older recordings for CD, and that spread naturally into mastering . . .

Soon after arriving at Abbey Road, I also started working for EMI's publishing arm, KPM Music, as an analog editor, splicing their music into 59- and 29-second versions for use in advertising. I'd just have a razor blade and a stop watch—it really trains you to listen hard before doing anything!

Q: Sorry, can I just ask one quick question? Those letters, were they basically just like, "Look here I am, here's my credentials, and I want to work in this field somehow"?

In those days, letters were all done on typewriters, as no one had a computer. I'd call up the company first and get the name of someone to type to. I'd tell them all about me, my musical background, working for BBC at the moment and looking to do anything in the studio world. I sent a couple of hundred of those letters. Back then you didn't get a great return on your efforts, but all that's needed is one interview. I had a talk with EMI's head of A&R in his really plush office in central London. After 15 minutes, he said, "I don't think you really want to be working for us. What you want to do is to be working up at the studios at Abbey Road." I said, "Well that'd be nice."

Weeks later I got a letter from Ken Townsend (the boss at Abbey Road) inviting me for a chat. Afterwards and he showed me around the building and editing rooms. This was six months after they switched from analog to digital editing. That had brand spanking new Sony 1100 and JVC editing consoles there, which they were all very excited about. He sat me down and said "I'm going to put it into play, and when the music gets to there in the score (it was a complex orchestral score by Stravinsky), you press that button." He was testing whether I could read music. Then the two of them went into the foyer and came back five minutes later and said, "Do you want a job?" I said, "Well thank you very much, that's very nice of you." That's how I got into Abbey Road. They got trouble from human resources, though. You can't hire on the spot. You have to advertise, even if it's only internal. Then all these people come around being interviewed for the job I already had, which is a little bit disconcerting. Nevertheless, I still kept it. One of the guys who did come was a roadie for EMI Records, Simon Rhodes. They reckoned he could do a few things. To keep him in the building, they put him in accounts for a while. He worked his way through, and he's now one of the top two or three film-scoring engineers in the world. He is about to do the new James Bond.

Q: So literacy, musical literacy, was an actual asset there?

Definitely. I don't think I would have got in without it.

Q: And Stravinsky scores aren't easy!

Exactly. His ballet scores are for very large orchestras, with complex rhythms. You need to be able to follow many musical lines simultaneously. Stravinsky's all over the place. As a kid, I was a bit geeky. One of my hobbies was collecting orchestral scores. I had a large collection of them. I'd sit with old LP records and follow the scores. As a result, I can follow scores very easily. I do a reasonable amount of film music. Before computer notation software, these scores were often A3 size, and sometimes even bigger. Often they would be written in shorthand. You just have to follow it without actually having to think about it. It's a skill you can't learn overnight. I spent about a year editing. If you know how classical editing works, it's very forensic.

Q: Sure; well, maybe could you take us through a little bit? Give us sort of broad contours.

Let's say you were recording a symphony, back in the analog days. There would only be one or two edits per minute as an average. You can't do many edits on analog tape. Once you've done it, you can't undo it. When digital came along, producers realized what you could do. You could literally edit out a semiquaver from a different take if you wanted to. The computer is your workstation. I suppose now the average number of edits per minute on a classical recording might be 12 per minute.

Q: That's fascinating for me to hear, as well. You often think, with classical recordings in particular, that there is a "live" element to it. The aesthetic to it is to make it sound as unedited as possible.

In fact, I know that there are plenty of classical musicians that go into a studio and record long takes. The assumption is that the long take is going to be used as such. I'm sure some of them would get a shock if they were told what actually happens to some of those takes. It actually isn't my decision; it stands with the producer. I just do what I'm told. When I first started editing, one of my first projects was the Chopin Etudes played by a very fine pianist. There were 1,600 takes for 60 minutes worth of music. That's a serious amount of takes. One of the last jobs I did before I left Abbey Road, around the late 1990s, was solo classical guitar, just one instrument. That ended up having about one edit per second. The review in the gramophone magazine said it had a "wonderful sense of style and phrasing." What the artist had done was play three or four bars, gone back two bars, play three or four more, and go back two bars, then play three or four more bars. You just wanted to curl up in a corner. Having said that, this is an extreme case. Personally, if I'm listening to classical recordings, I prefer to listen to old ones, because I know that there isn't so much editing and post-production going on. This doesn't mean that the artists were better back then at all, just that the recordings are more natural, for want of a better word. Even with live recordings, it's amazing what you can do. Normally, a big label that is doing a live recording will record three concerts, one after another of the same programme, bang bang bang. There will be rehearsals before and possibly a session for patches afterwards. You can do an infinite amount of editing with all that if you want to. It is live, but it is edited.

Q: It's aesthetically live.

That's right, it's aesthetically live. The problem is that everyone knows if there's a bummed note or something wrong with the timing, you can correct it. You can correct in such a way that people are not going to notice. So there you go, what do you do? That's editing for you.

Q: Yes, thank you, that's interesting.

Isn't it? The only format which was never edited was 78s. You stick on a 78 and you know it's not been messed around with. You get what you get. If they messed up on the record, they chucked it in the bin, put a new one on the platter and started again.

Q: So getting back to mastering, then, just about you getting a gig . . .

Yes, that all started with remastering, really. I started at Abbey Road during the early days of CD, and all of EMI's back catalogue needed transferring onto the new format. That's how I got into that world. After a year of editing, I knew I had to branch out, and the bosses needed people with musical knowledge to deal with these tapes. I must have done this for five or six years before the first real mastering came my way, and by then I'd really learned how to listen to sound, realizing when there's a problem, how to EQ, and so on.

Q: So you were saying that you had to do the transfers from tape to compact disc and there were no rules . . .

It was quite interesting, because there were no rules. No one had ever done it before. We were just given the analog tapes and were told, "We need a CD re-master of this," and away you go. That's one of the strange things about being in a studio back then. No one told you what to do or how to do it. It was very strange, really. We came up with all sorts of ways to correct faults such as audible edits, dropouts, print-through, rumble, etc. How to EQ or not was also left to us. Mind you, we did have lessons in lining up tapes, playback levels, and all that technical stuff. It was fabulous, really, and a big responsibility, too.

Q: Can I ask you a couple questions just about that transfer process? When you were transferring, would you find yourself rebalancing, playing with the dynamics, or did you find yourself an optimal level for transfer and transfer edit, for instance?

Back in those days, the number one thing was getting the right level off the tape—making sure the tape machine was properly lined up and everything was paramount. We didn't have any choices as to what converters we used. Back then there was only the Sony 1610 and 1630. Then it moved to other things like Genesis or Prisms. It's just getting the right signal off the tape through the best convertor that you have access to. Once we've done that, we messed around with EQ like you wouldn't believe. The first things were sorting out all the problems that weren't picked up when it was actually recorded. You get ground loops and main hums. You get harmonics over here that would be 50 Hz, and 60 Hz in the States. Often it would go up the harmonics of 100, 150, 200 Hz. The Urei "little dippers" were brilliant at dealing with this. Then there's rumble, which can sound dreadful on a CD. We'd play the tape whilst following a score of the music, and roll off the bass end when bass instruments weren't playing. Regarding the HF, we'd sometimes ease it off a little if the hiss was obtrusive, but personally I like to leave this alone and not risk pulling down some harmonics, making the sound duller. From 1968, Dolby noise reduction was used on most recordings, making the background noise far lower. The rest of any EQ was down to experience and knowing whether the frequency is even across the spectrum. We never used compressors, but sometimes did some manual dynamic control just using the faders on the mixing desk. Dropouts could be a big problem; we did amazing things with ambience and reverb to try to get rid of them. It's easy now, with computers, but back in the 1990s we were inventing tricks all the time.

Q: To accommodate for the dropouts?

Just have reverb up and the fader to hide it a bit. You can get away with these things on an LP, but you can't on a CD, especially with headphones on. And as far as the overall sound goes, a lot of the time you leave it alone. You only change if you have to; the trick of course is knowing if the sound needs work or not. Quite a lot of these early tapes tend to have a bit of a boost around 2.5 kHz. It was a deliberate frequency response in the early Neumann microphones to give a bit more of a presence boost. Sometimes it can make it sound a bit harsher. Quite often, we dip in that sort of area to get more oomph in the bass line. That's not so crucial in classical music compared to rock for bass.

Q: It was corrective, I suppose? You might call it that.

Yes, it was corrective, and also making sure all the music was there. It's amazing how often it isn't all there, and you really don't want a recording released with something missing! There is a huge library of music at Abbey Road. You were asking for trouble while doing this work if you couldn't read music, or you had to pass it on to a third party. Back in those days, there was quality control. They would listen to the master which you had created to make sure everything was okay before it went off to the factory.

Q: I imagine that would actually be reassuring in the early days, or intimidating.

Back then you took quality control for granted. I think we'd have freaked out at the thought of sending masters off without someone else to check them. These people listened to it and put all the PQ information through.

Q: Oh, I see. So that's sort of a role that's been absorbed by mastering in general now; if you're an engineer you're supposed to be able to do all of it now.

Yes, and of course that side doesn't take very long to do.

Q: Don't tell clients that!

I wouldn't mind if all clients listened through it before sending it off.

Q: I actually like that idea. As long as they know what they're listening to.

I just make sure that whoever the client is knows that they have to check it. If they don't, then that's their problem. I don't want to be sued.

Q: So was this like a 9 to 5? You would come in, you would have this library of tapes that need transferring, then do what needs to be done? Or was there some sort of order to this? How was the brief given to you during this work?

In those days, you charged for the amount of time it took to do the job. There was an hourly rate. Money wasn't so much of an issue in the 1980s and '90s. All of the post-production rooms in Abbey Road had two people in them on a shift basis. It was quite flexible. Management didn't care what hours you did, so long as the work got done. There were all sorts of shifts going on, and you could split it in half and half each day. Start at 8 in the morning until 3 in the afternoon, so long as you added up to about 37 hours a week. They wouldn't mind. Some weeks had three 12-hour days with four days off. As for prioritizing the work, we were often overrun and did it on a "who's shouting loudest" basis!

Q: So you're doing this, and I assume that you are slowly becoming identified as a mastering engineer, someone who masters in this field. Would you say that's the case? You become known through this work.

I did become identified as somebody who works specifically in that area of music. I'm perfectly capable of doing other genres, but you get shoeboxed. People who are looking for that work find you. Other artists in different genres, like Pink Floyd, often don't find out about you, though you're perfectly capable of doing that type of work. You're bound to get other stuff from time to time at somewhere like Abbey Road. In the early days, I walked in and spent the day with Pet Shop Boys. I didn't know I was going to do that until I walked in the front door. Same with the Red Hot Chili Peppers, Paul McCartney, and others.

Q: I suppose that's the magic of working at Abbey Road?

You never know who you're going meet walking down the corridor. You get top-class artists when you're in a place like that. You do get to see some interesting things. When Oasis were in studio 2, we were all curious to see the setup. They must have had 100 guitars along the wall of the studio (maybe not 100, but lots!). One day in the restaurant I was stood behind Leonard Bernstein, and in the corner was Kate Bush. You don't get that in many places!

Q: You never know.

You do get shoehorned a bit, which can get frustrating. That would be another reason why I decide to move away and do my own thing.

Q: Okay; can you talk a little bit more about that, the transition from this sort of start at Abbey Road to becoming, I don't know if the right word is freelance, I mean I'm not sure how you describe yourself?

I started my own company. I'm employed by my own company, which is sort of freelance. There were a number of reasons why I decided to move. One reason was that we had little kids, and London isn't the best place to bring up a family. I was starting to get frustrated at Abbey Road because I was always doing the same thing day in and day out. I've been there for 14 years. You couldn't move sideways into recording because there were other people doing that. You would stand on toes. You got brilliant at not doing very much. At the end of the day, it's very frustrating. The catalyst was one weekend when we were in Yorkshire seeing what property was available at the price we could afford. My father was an architect, and he said, "Come have a look at this house." We said, "Don't be stupid," but he said, "Go have a look, just see what it's like." We did, and ended up buying it. We put in a silly offer, and it got accepted. We said, "Oh my God, what are we going to do now?" I suppose if you think too hard about these things you don't do them. It was almost an accident.

Q: So you bought the house, and you said I guess I got to move?

That's sort of how it happened.

Q: Can I ask you a few questions, then, of how you set up shop? Do you know where you would work? Do you have your own space? Did you have designs? Where were you when you left Abbey Road, when they had all this fabulous equipment?

One of the reasons we bought this house because of a big, 25-foot-long attic. It actually sounds quite good in there. First thing we did was convert the attic into a suitable place to work in. We put a floating floor in it and acoustically treated it. The other thing that made it possible was that I got a contract to work on *Captain Corelli's Mandolin* in 2000. Financially, that made it work. As far as gear goes, I cloned my room at Abbey Road. I just got exactly the same gear apart from ATC loudspeakers. At that time, the DAW everyone was using was Sonic Solutions. When mine died, I switched to Pyramix.

Q: Are you still on Pyramix?

Yes, I am, a couple of versions back from where they are at the moment. I'm still on an older version, but it does what I need it to do. What's the point of spending all that money if I don't really need it? Mastering the sort of work I do, you don't need 128 tracks.

Q: If you know it and you're comfortable on it and you're not missing anything.

It's got one or two irritating things on my version, but I've got workarounds.

Q: And now that's part of your workflow, right?

It is as it is. They wouldn't allow third-party plugins on mix buses. On the strips, you couldn't automate them. You could automate their stuff, but not third-party stuff. However, it forces you to think up alternative workflows and ways of doing things.

Q: In terms of mastering, have you noticed any sort of changes in what you're working on?

Sometimes you get projects in to master which don't sound very good. They need more remedial work. Either that or you get back to the client and say this, this, and this is wrong. You ask them to have another go at it and send it back again. If I can fix it, I will, but sometimes it is best to go back to the mix. Sometimes levels within tracks or EQs are not right because people are mixing in less-than-ideal spaces. It doesn't matter how good your ear is—if the sound in the room is wrong, you're going to EQ it wrong.

Q: Are you finding, this is a question we are asking, are you finding that you are doing this more now? You are working with clients that probably have much better environments. Just in general, do you find that you encounter the need to do this more, those sorts of sessions?

Yes, I think so. Lots of areas can be less than ideal if your working environment has issues. Quite often there are phase issues, probably because some people aren't aware how to avoid this right back at the tracking stage. Sometimes the imaging isn't right—too narrow, or a hole in the middle. The panning is sometimes wrong. In a classical recording, generally all the bass and cello stuff is off to the right.

Q: Mimicking the actual distribution of an orchestra.

What you are trying to do is recreate what was in the orchestra. You're not trying to do anything fancy or creative. You're just trying to get what was there at the time it was recorded, a natural effect. Even with classical music, I'm noticing it happening more. Things come in to be mastered, and they need a fair bit of work done on them. On the flip side, you can't charge as much money as when I left Abbey Road. The same work now, 15 years later, you're getting 20% less money for, which is an issue.

Q: Absolutely, that's part of it. So you would recognize that as a trend?

There's so much music around these days. So many people think music is for free. I'm in this part of the world where people are buying fewer and fewer CDs. I keep saying to my main clients, "Why don't you stream stuff, or have it available for download on iTunes? You only make CDs." And the client keeps saying, "No one's going to buy those, are they? My clients at least, the people that buy my stuff, want the hard physical product. They're not into computers, so we just sell CDs." Sales are dropping all the time.

Q: Why might you run an analog session? Do you find that clients still care? Do you find that clients care what gear you use?

Most people aren't interested in analog. These days it's expensive, cumbersome, and not as flexible as digital. I did produce a recording at Abbey Road that was specifically to be recorded onto half-inch analog tape running at 30 ips. People are still around who want to do that. Half-inch tape at 30 ips sounds totally fantastic. In the film world, there are still various projects that get backed up onto analog tape. You know that in 50 years' time you'll be able to play it. Who knows: if you have a hard drive or another format, will you be able to play it in 10 years? There are people who are worried in that respect of digital as to what's going to happen. I'm doing analogue tapes all the time from 1949 to 1950, and they're still playable. There might be a few dropouts, and a bit brittle in parts, but you can mend the edits. They still play. I don't think analog will disappear—the sound can be amazing, and there will always be people who swear by it and will pay for it. It's the top end of the market, though.

Q: Yeah, that's amazing, because that runs counter to what I think popular wisdom says concerning the formats there.

I think it probably does. I think analog is still here to stay. One of my students at Leeds College of Music is now working in Nashville in a purely analog studio. They are up to their neck in work.

Q: What's the studio? Do you know?

I knew you were going to say that. I think it's Blackbird.

Q: Thanks. So you mentioned mid/side work. Do you have any sort of modus operandi? Do you mono lows, for instance? Do you mono the lows in classical? Do you even do that?

I'm doing a job at the moment which will be broadcast on the radio on classic FM. Obviously, you want mono compatibility. I'm checking it all the time with a phase meter, playing sections in mono to see if it sounds all right. I know it's going to just be played on a stereo and mono systems. A lot of the time, I'm not too worried about it. There's no point in mono-ing lows in classical, but it's worth it in popular genres. Also, you can use mid/side to roll off low frequencies away from the centre, creating a cleaner sound, and more space, too.

The more stuff I do with the students, the more I'm making myself acutely aware of it. When something sounds a bit phasey or something, I notice it more than I used to.

Q: Why do you think that is?

I think it's because I'm starting to mark students' work in about a week's time. I just know there is going to be phase issues on some of it. You get tuned into it in a way that I wasn't before, listening to stuff that was professionally done at Abbey Road. It wasn't so much of an issue. When you get outside of that environment, a lot of people don't think about things like phase. If you have phase issues and you do hit that mono button, it can sound like absolute rubbish.

Q: Yes, and so if you're willing to talk about it a little bit, what are some things you do correctively?

The phase can be improved in various ways. You can bring in the image a bit from left to right and make the whole thing a bit narrower. You can sometimes help if you split it into mid/side and get rid of rubbish on the left and the right. Make sure the centre is all happening nicely. I confess to being a bit naughty on a few occasions with sections within a track going out of phase into negative. I have to double patch and flip the phase 180 and plug it back in.

You have to be a bit careful at the two edit points, but I have done that. I can't think of anything else I do off the top of my head.

Q: In terms of metering, do you peg at a particular value? How do you meter?

I have the VU meters up on Pyramix. If you're doing classical music, you want dynamic range. The view I take, if it's full on big orchestral things, it's going to be peaking close to zero. Or whatever new delivery medium we're working to these days. The main thing for me is the monitor control. I spend some time playing different recordings and finding where I like the monitor level to be, and then I leave it well alone. That way I know that anything I work on is going to have the same relative level and not go over peak level. I'm much more interested in the average level rather than peak, and love to keep lots of dynamic range. If an album is all quiet music, why have it peaking close to zero? This way the listener at home doesn't have to change the volume level between albums.

Q: So you peg your monitors, then?

I'm always monitoring at the same level. I don't touch that. If I want to do that, I do it on the faders. I don't change my master output level. I'm always in the same ballpark. I think it's quite important, personally. This thing about dynamic range is everyone's saying you don't get enough. Sometimes it can go the other way in the classical music world. I was doing a recording of a larger symphony with an EMI producer and the whole thing starts soft, with the double basses playing pianissimo. The way they were playing it was −50 dBs of headroom. The producer is sitting there saying, "Fantastic, you got to listen really hard to hear it, isn't that fantastic?" Then, of course, three minutes later, BANG. If you set your levels for the opening, you're either going to blow up your speakers or you're going to blow up your ears. You've got to have a happy medium.

Q: That's excellent. That's such a great counterpoint to the kind of invective around the loudness wars that's happening right now.

I don't see the point of it. You have 24 and 32 bit recordings, and all this headroom. Why only have 4 or 5 of dBs used in a recording? I'm sure you heard that before.

Q: Do you do a lot of stem mastering? Do you do stem mastering?

I have done stem mastering. It doesn't happen very often, but I have. I don't mind doing it. It almost seems like a cop-out from the mixing engineers' point of view. It's like they can't quite make a decision. It's fine if something specific needs to be done to one of their stems and not the other stems. If they are being specific about their request, then that's cool. Then you think: something so specific to that stem, why don't you do it?

Q: Do you charge more for stem mastering?

I would try to. Whether I get away with it or not is another matter entirely. I normally end up agreeing a price before the job starts, and we have this to and fro bartering about money. Almost everything I do is fixed price. I start the job, and that's it. If you got it wrong, then you take it on the chin. For my main clients, it's a sort of a fixed price per job. Some jobs are quick and some jobs take longer. It has its roundabouts, but evens out over time. We're all cool with that. If it's taking a bit longer, then that's fine, so long as you do the job properly. I hate not going the extra mile—it's pride in the job, isn't it.

Q: Where do you think mastering is headed?

I know people who used to be mastering engineers and aren't anymore. The amount of work seems to be going down. The top-end stuff is still going to keep going because people at that end understand the value of mastering. People in the middle area may understand the value but don't have the money. Some think they can do it themselves. I'm convinced of that. It's tricky; I'm sitting on the fence at the moment on this one. I'm not quite sure where it's going to go. I think there's always going to be a market for us as mastering engineers. It's heading to the top end of the industry. The big names and the big bands are quite happy to spend the money and get it right. In the classical world, mastering has never really existed. What tends to happen on a classical record is, traditionally, they will spend the first hour of a session getting the sound right. Typically, they move microphones, listen in the control room, make comments, make adjustments, do another take, have a listen to that, get the sound right, and then track the session. In post-production, you don't change it unless you're told to, since everyone on sessions has agreed to the sound. Traditionally, that's how classical works. Maybe you might bring up the whole levels to peak at the right amount of headroom. Mastering doesn't really exist in the classical world. It is really only in the "pop" world. What I would say is that those people doing their own mastering work, at whatever level, should educate themselves. Seek out advice and tips from people with some experience. Even better, put aside a portion of your budget for it, because it's important.

Q: What we're hearing is that there really isn't clarity right now.

Yes, I don't think there is. No one really knows.

Q: The wheel is still in spin.

I think what we need is less music out there. If the music in shops, elevators, and lifts wasn't there, then people would be more receptive to music. I think this is why vinyl is coming back. It makes music an event again. You have to sit down and actually make an effort to listen to it rather than just in your headphones walking down the street.

If vinyl is coming back, then mastering will have to come back in some way. Whether it's long term or not, who knows? I don't know if it's a fad or if it will be long term. I've got 20-year-old students at buying vinyl all the time.

Q: Where do you store it?

Why are you buying it? It's more expensive. I think people love vinyl for the physical size of it, and the artwork. I just had a student only last week explain this to me. He said "You probably don't get this, because you've been doing a load of analog in your career. For me," he said, "I love the noise. I love that sort of crackle."

Q: And of course when compact disc came out, it was no more surface noise.

Exactly; it's quite funny, isn't it. So yes, there you are.

Dr. Phil Harding

Phil Harding joined the music industry at the Marquee Studios in 1973, engineering for the likes of The Clash, Killing Joke, Toyah, and Matt Bianco by the late 1970s. In the 1980s, Phil mixed for Stock, Aitken & Waterman, tracks such as "You Spin Me Round" by Dead or Alive, followed by records for Mel & Kim, Bananarama, Rick Astley, Depeche Mode, Erasure, Pet Shop Boys, and Kylie Minogue. In the 1990s, Phil set up his own facility at The Strongroom with Ian Curnow. Further hits followed with productions for East 17 (including "Stay Another Day"), Deuce, Boyzone, 911, and Let Loose.

Recent projects include the book *PWL from the Factory Floor* (2010, Cherry Red Books) and mixing Sir Cliff Richard's 2011 album *Soulicious*. Harding has recently worked for Holly Johnson (Frankie Goes To Hollywood), Tina Charles, Samantha Fox, Belinda Carlisle, and Curiosity with his new production team PJS Productions. Phil is also vice chairman of

JAMES (Joint Audio Media Education Services) and was the chairman of the Music Producers Guild for many years. Phil completed his doctorate at Leeds Beckett University, April 2017.

Q: How did you learn about the existence of mastering?

I started at The Marquee Studios as a tape-op/assistant engineer in 1973 but did not start engineering records until around 1977/78. The first record that I mixed (rather than just recorded) and was invited to attend the mastering session was a single for Long John Baldry, produced by Jimmy Horowitz. The mastering session was at Trident Studios, just around the corner from The Marquee, and I think the mastering engineer was Ray Staff. My lasting memory was that I couldn't understand why the dramatic finale on the record was causing distortion at the level of loudness that the producer wanted the overall 7 inch to be. I would later come to understand this fully when I began to master albums I had engineered and mixed with Tim Young at CBS Studios.

Q: What sorts of things do you expect mastering engineers to do now, or in the past?

I think now and in the past, the main thing a mix engineer and producer expects the mastering engineer to do is to give an honest opinion of the track(s) in terms of the overall sound, the requirements needed to make the running order work well and potential processing both overall and for each track. This can take some time and multiple listens in the mastering room, and there is every chance that I will formulate some different views during that listening experience, being in a different room to where I have mixed. Ideally, though, it's the "fresh" view of the mastering engineer who is hearing the music for the first time that should be noted and acted upon first.

Q: How do you accommodate/account for (make room for) mastering in your production process?

I will always allow the mastering engineer to speak first after the first listening session, and I will always respect that they know their room and speakers at a much deeper level than myself. I think engineers and producers need to trust the view of how the mastering engineer feels the track (after their processing) will sound outside the mastering room. Importantly, though, the attended mastering session is a collaborative process where the client and mastering engineer need to respect each other's opinions and views of processing, suggestions coming from both sides.

How has mastering changed, in your experience, over the course of your career?

It is easy to say immediately that there was a huge shift from vinyl mastering to digital mastering. The most significant thing at that time was that suddenly one could have loud tracks all the way through an album without having to compromise with levels on the inside tracks to a vinyl album on side A and then side B. This problem would often affect the producer's decision of the running order. The other significant thing to my mind was/ is that sibilance is no longer the problem it was compared to the analog/ vinyl era. Too much sibilance on vocals would distort very easily during analog mastering, and for myself, working mainly in the pop music genre with loud vocals, this would often cause mastering problems. The trick became to be sure that, as the mix engineer, there would always be de-essers at the end of the processing chain of the lead vocals on the mix. The most embarrassing thing in the late 1970s and throughout the 1980s was for a mix engineer having to witness mastering engineers applying de-essers at the mastering stage because you had not done your job on the mix session. This was very detrimental to the overall mix, in my view, as de-essers at the mastering stage can also affect cymbals and other high frequency signals within a track.

Q: Do you expect different things of mastering now than you once did? Why?

What I hope for now is the opportunity for the mastering engineer to step into the project in the same way that producers and engineers have to now, with an expectancy that artists, managers, and labels will change their minds and request multiple mixes with contrasting suggestions and therefore multiple mastering sessions, as well. This is because everybody now knows that we can instantly recall our mix and mastering sessions on our computers without many hours of analog equipment resetting. This may not be quite as easy for mastering engineers and mix engineers still using outboard analog hardware, but it is easy if everything is "in the box." This attitude requires both business and creative flexibility on the part of mastering engineers, which I believe the mastering world is embracing in the same way that producers and mix engineers have had to.

Q: Do you have a favourite mastering engineer(s) to work with?

I have been lucky to work with some great mastering engineers over the years, and my favourites are Tim Young, Richard Dowling and, more recently, Nigel Palmer. The reasons are often as simple as the convenience of being able to attend a session that is nearby to where one works and the trust that is built in having a mastering engineer working on your material multiple times with excellent results that are both accepted by the clients and may even become chart hits.

Q: Where do you think mastering alters and adapts your work? When might this change?

It is only in recent times that mastering has altered or affected my work to the point where I will go back into my studio to make adjustments as opposed to making those adjustments in the mastering room. The ability to recall a mix instantly and alter individual elements of the mix is a new phenomenon in the relationship between mix engineer and a mastering engineer. File sharing systems such as Dropbox have made this flexibility possible.

Q: Do you think you produce (or engineer) differently based on your musical outcome, and thus the format it ends up in (say a segued concept album, or a particular delivery method such as vinyl only, for example)?

Yes, but not to a major degree. I have not mixed any projects in recent years that have been heading for vinyl mastering, so none of those constraints that I have mentioned come into my thinking these days during a mix. The most significant change I have experienced recently is that I have worked on so many individual tracks for singles that one loses the concept as a producer and mix engineer of listening to the running order of an EP or album before attending the mastering session to be sure that there is a good compatibility of overall balances from track to track. I have recently experienced this whilst mastering an EP, where I had not performed this task in my own studio, only to discover a mismatch of bass guitar levels between tracks whilst in the mastering studio. This is something that I would have checked in the past, but we seem to become so focused on one track at a time now, due to living in an age where listeners purchase and listen to only one track from an EP or album to add to their playlist compilations, and industry practitioners get drawn into this same trap of losing the focus of a whole project's multiple-track running order.

Bruno Fazenda

Bruno Fazenda is a senior lecturer in audio technology at the University of Salford. He graduated in 2004 with a PhD in room acoustics and psychoacoustics from the University of Salford. After that he was a research fellow with a Marie Curie research fellowship at the Danish Technical University, and has been lecturing in the UK since 2005.

Bruno is now part of a cross-disciplinary EPSRC funded network focusing on Acoustics and Music of British Prehistory. His work on the acoustics of Stonehenge has been reported in an article in the *New Scientist*, presented in a documentary on The History Channel, and featured on news and information sites, such as the BBC and *The Guardian*.

Bruno is a member of the Institute of Acoustics, the Audio Engineering Society, the CIKTN (Creative Industries Knowledge Transfer Network), and Immerse UK. He is on the technical committee for "Perception and Subjective Evaluation of Audio Signals" for the Audio Engineering Society. He has been invited to present special lectures for the Institute of Engineering and Technology, the Institute of Acoustics, and the Audio Engineering Society.

Q: How did you get into this kind of research?

I'm a senior lecturer in audio production, which means that I deal with a number of aspects, mainly in teaching, to do with producing audio, recording, audio post-production, and a little bit of mastering and mixing and so on. My research is slightly detached from that, in a sense that I'm interested in the subjective quality of audio, the way people perceive quality in audio signals. A lot of my research has been looking at metrics that can quantify the quality in a signal, and these would be things that we can extract from the signal directly, without asking anyone any questions, then trying to correlate those to how much people tell us they like a particular song, or, at the moment, we're working on various mixes of the same song, and then trying to find out of all of those metrics, which I call "objective metrics," which ones are relevant to basically tell us that we're in the presence of a good mix or a bad mix, or a good master or a bad master—that's basically it.

I have done some studies on the quality on the acoustics in rooms and started going into the quality of signals. We've done a lot of projects at Salford University that look at this from various aspects. One of the aspects is, of course, the way in which signal parameters can tell us about signal quality, but I'm also very interested in aspects to do with when you ask someone, "Is this good quality?" What do they perceive from that question, and do they understand the kind of answer that they're being asked for? So we've looked at things that compare, for instance, how much someone likes a song, perhaps from a more hedonic, or more personal and emotional taste, or how good the quality of that song is? We found that, for instance, if you asked those two questions, then the answer to the quality question is slightly different compared to if you only ask about quality. There are some aspects that go a little bit beyond just the signal itself and just focus on more of the psychology of the way people think and the way you can actually get responses from them.

Q: Are there any automatic things that you're finding from your research that would go onto informing, say, the way a mastering engineer views sound? For example, a lot of what we've asked mastering engineers is how do they know when something sounds wrong or

right. Have you got any connects between the research, the interested stuff that the listener would say was great-sounding music because they enjoy it, or the subjective sound quality? Is there anything that connects the two at the moment?

I understand, yes. With Trevor Cox, for instance, we have recently finished the project that we are now working on what we would call a "Global Meter," a global sound quality meter. It's a particular one, because we weren't looking at commercial music, we were looking at user recordings, things that people do with their own devices such as cameras and mobile phones. What we did was that we did a whole lot of studies where we came up with machines that could detect the errors, we called them "errors" in this particular project, and we've looked at three main errors—we looked at distortion, wind noise, and handling noise—so these could be used for things like microphones in recordings and so on. We've also done an extensive amount of subjective testing where we've asked people to tell us their perception of the quality of each of these recordings. Then we came up with some machines that could go through unknown and unseen samples which contained these "errors" and award them a certain quality rating automatically. We then compared how that machine was doing against people rating exactly the same unknown samples, and we found a pretty good agreement. One of the things we're moving towards now is to put together a paper which shows the final Global Meter. It's basically something that can look at a file and tell you from a pretty safe perceptual point of view, how well that file will be received by listeners. Now, that is mainly looking at aspects to do with recording errors, so it doesn't apply directly to music, but on a different project with a PhD student, I am looking at exactly the same ideas, but now looking at recorded music parameters. So, we've got a number of algorithms that can extract these features and pull them together and come out with some sort of quality rating, but we haven't kind of finalized the model for that yet. So at the moment we're investigating what are the relevant parameters? What are those that make more impact in terms of people's assessment? And what are those that we could probably leave out of the model, because they're not as important? As a final sort of answer to your question, those models and automated processes do exist—they exist now from our research for things to do with user recordings—but I'm quite sure, within the next decade or so, we'll start seeing a lot of meters that report quality rather than just, say, loudness, or frequency or whatever.

Q: I was wondering if you could go into more specifics in terms of "sound" lacking in quality that your error meters were bringing up. Were there any broad patterns there?

What our models show is really the relative importance; for instance, in this case, when we're talking about recording errors, they talk about the relative importance of these, and the interesting thing is that they're not surprising at all, but they're scientifically proven now, so things like distortion and frequency range are seen much more important in their impact to quality than things like handling noise or wind noise. So the relative importance of these errors is higher for distortion and high frequency. I'll have to get back to you on that, but I believe I am quoting the right ones.

Then when we look at quality in recordings, then the important or main components that are coming out, again not surprising, the most important component is loudness. Loudness is really one of the key things that mark perceived quality, but it's not loudness in terms of how particularly loud something sounds, because we normalize for that, so when we do our tests, nothing is louder than anything else. So this result is showing that the important bit is more to do with dynamic range, which is still a quantity which has to do with loudness, or in some cases loudness range, so that component is always coming out as the most important. After that it's frequency range again, so wider frequency ranges always sound better than narrower ones. Then beyond that, with a lot less importance, the people are telling us, or the results are showing, that it's things to do with panoramics, with stereo width, for instance. We've not studied anything in surround, so we've only looked at aspects to do with stereo, and finally the bass response. These are the important metrics, let's say, that are more directly correlated to quality.

Q: So, when you're saying that you're "finding that louder is better; however, all the tracks are normalized," we're assuming you mean peak normalized—so are we talking about average amplitude for loudness?

No, I need to correct you on that. I said that loudness was an important parameter, but, and again I don't think this is a surprise, it's inversely correlated to quality. So what we're seeing is that dynamic range—higher dynamic range—is higher quality. So I guess that falls in line with a lot of what's being discussed in the press, or in particular press for audio, that says that quality has been decreasing: we're finding the same thing. What we're finding is that the recordings, which show a much narrower dynamic range, are the ones that are rated worst. So, the component is inversely related to quality.

Q: Does your research at all touch on these sort of social ideas about loudness, and how that might impact the evaluations of loudness in your listeners?

I think the direct answer to the question is very simple, and that is, that if you compare like for like, and one of those has just one difference in that it is 1 or more dB louder, then that will always be preferred. So, loudness does matter; it always kind of overrides most of the other aspects, until you hit some pretty horrendous problems, but anything that is presented at the louder level—up to a certain limit, I would say—but anything that sounds louder will always be preferred, and I think that's where the whole thing started. Now the social aspect of it, or let's say the technical social aspect, I think, is down to trends that have started sometime around the '80s, where a lot of this music was being broadcast on the radio, and the guys in the radio realized that they could actually maximize, let's say, the impact of their music by processing it, and the simplest process was to make it louder. Of course, you guys would know that if you want to make it louder, you compress it, and then you push the level up. So there were these processors used in radio broadcast that found their way from the radio into studios, and people started mastering their final products so that they were coming out of the studio louder. After that, it was a race: everyone was trying to get it louder and louder and louder. I think in some ways—we can see this from the state of the industry—people just lost control, and just kept on compressing more and more and pushing the level higher and higher.

It's important to see that the more you normalize something, the more you push it up, the louder it's going to sound. So if people are thinking, "Well, if it sounds louder, it's going to sound better," then of course they're just going to push that to a limit. What we're finding in research is that that's not necessarily the case. If you remove the loudness aspect and normalize everything to the same level, then people actually prefer the higher dynamic range, compared to the very squashed dynamic range.

To answer the question you've asked before: when we normalize according to loudness, we use one of the standards. It's not peak loudness, it's more akin to RMS, but it's perceptually based: it's an algorithm that goes through the portion of signal that you give it, and it will tell you that it's got a certain loudness in dB.

Q: What's your views about the accessibility of digital technology, and obviously no longer having headroom?

I have a personal view on the digital technology, and when that starts finding its way into the bedroom studio, the idea now that anyone can make their own music and of course then publish their own music. I think there is sort of a trend that appeared there, where amateur producers or musicians became a hybrid of a musician who then produces their own music, and these people were obviously looking around for inspiration, or for guides on how to publish their music. Of course, one of the first things that you realize is that "Commercial music sounds loud, it sounds louder than my own bedroom recordings. So what do I have to do to make it sound at least as loud?" That's the point where you start using your Ultra-Maximizers, and any other processors that did the same thing, to start pushing the levels up. Of course, the inevitable outcome of that is that you start squashing the dynamic range, and a lot of it happened through people not knowing what they were doing with the gear, and I include myself in that group. In the '90s, I was doing my own bedroom stuff with friends and going back and trying to mix it. I was supposed to be the sound engineer, and I was finding myself saying, "Oh, this is an excellent tool! I pull this thing down by 12 dB, and all of a sudden it's much, much louder!" But of course, a lot of people were doing this oblivious to the other effects that they were having on the actual music quality. Beyond that, I think there's a slightly more obscure progression from that, although I haven't found any evidence for it, but I have a feeling that it might be happening as well, which is as you start pushing these levels up, as you start squashing the dynamic range down, you start getting this more "crunchy" type of output from it; you also start associating that with "pro level," thinking, "This is how it should sound." So you almost accept those small amounts of distortion that start appearing as a kind of "mark" of "produced sound," and I think that's where the whole thing starts sliding down, the whole quality perception starts changing, because people stopped realizing that, actually, good dynamic range and clarity and lack of distortion actually makes for better sound. It's almost like people became habituated to this kind of really crunchy, very compressed sound, and there are a number of artists that would serve as good examples. This just became the norm without people realizing that the quality has gotten lost.

In the digital domain, if you hit the maximum bit, then you're basically squaring the wave and creating some really nasty distortion, whereas in the times of analog distortion was a bit more benevolent, in the sense that it was gradual. Of course, if you were to distort it very drastically, then you would still get the nasty effects, but I think that digital is a little bit more precise in a bad way in this case, so as soon as you hit the maximum of the system, then the wave gets squared off entirely, whereas with analog, there's a much more gradual and harmonic type of distortion being added, so it was much more acceptable in terms of perceptual effects.

Q: Have you got any research that backs up the following: If we were to take a sample of records through the 1950s through to today and beyond, each decade, to some degree or another, all have a different EQ curve and a different harmonic response as you pass through the various decades and trends. Have you thought about that, and are there any preferred qualities from the people you have asked about this?

Most of the research out there is, again, based on metrics, quantities that you can extract from the signal. Of course, it becomes a little bit more difficult to do that research earlier in time, because most of that was analog originally, and in order to analyze it with the tools that people have been using, digital tools, digital algorithms, so they had to be converted from analog to digital. Now if that conversion was done directly, then there's no problem, but what you find is that a lot of the research sometimes goes onto mastered or remastered versions, and of course one doesn't know what happened at the mastering stage and whether some of the mastering tools applied confound the effect because they now add modern stuff. So if someone picks up something from the '60s and masters it, they might actually maximize it just like they would any of the modern records.

With regards to your question directly, there is some research out there, and I think it would be more credible from the digital era onwards. It shows a number of things. It shows, for instance, a steady increase of loudness between 1965 and 2013, which is well documented in papers. I think loudness started off at below −10 dB, and it's now verging onto up to −5 dB, so there's been a 5–7 dB increase in loudness. Obviously, this depends on how you measure loudness in that sense; there are a number of algorithms, as I mentioned. There's a very clear reduction of dynamic range, again because it's related. You've seemed to have lost at least something like 5 dB of dynamic range on average between 1965 and 2013. Then there's a whole number of different metrics that show some of this progression over the years. So things like the Peak to RMS Regression Coefficient, which one of the authors puts forward as one of the best ways of actually measuring what's going on between all of the years, that's well documented.

More recently, one of the things that I've done with one of my PhD students at the moment, Alex Wilson, looking at quality in music signals corroborates more or less the evidence that we were talking about. We found that quality is very strongly correlated to this dimension, to this component, which has to do with metrics of loudness and dynamic range. When we display our music samples of our analysis over this dimension, and we look at how they group in terms of decades, we clearly see—and again, our analysis only goes back to the 1980s because we were only using freshly released, commercial digital music, so we didn't look at any of the analog remastered stuff. What we can see is that it starts off at the highest perceived quality in the 1980s, then it moves down in quality in the 1990s, and then the 2000s and the 2010s sort of occupy the same region in that dimension. So this is basically saying, 1980s starts with the top perceived quality, and then as you go forwards from that, quality starts decreasing. This data correlates our perceived quality by our test subjects with our analysis of commercial music. but then if you look at any of the other people looking at this, you also see a kind of progression like this, which doesn't have the quality aspect because these people were not doing the subjective testing side of things, but it clearly shows the dynamic range being completely squashed, and it clearly shows features of the signal being changed over the years. So, I guess that answers a little bit of your question, that there is evidence that things have been changing over the decades.

Q: Where are you planning on taking these sorts of studies? Are they purely academic, or are there commercial outputs?

Well, I work very closely with a guy from Queen Mary University in London, Josh Reiss, who I would say is an expert in automated mixing. So our paths crossed because he has been doing a lot of work with getting many tracks and mixing them down automatically with no human intervention. But of course, he then faces the problem of, "Well, what sort of criteria do I use to mix this?" Of course, his work has been based on what we call "best practices," such as, "We want all tracks to be balanced across level, and then we want vocals 3 dB louder," or something like that, so he's got a number of points to start from. But we don't have any data, or any kind of scientific data, that tells us: "We want dynamic range to be 'this' level, we want the bass to have 'this' kind of dynamic range and to be at 'this' level, we want drums to be compressed like 'this,' panned like 'that.'" So, a lot of this work was to try and find these criteria, the subjectively relevant criteria to help automatic machines to mix down and to get to better quality. We are now moving towards having, I think, a more modern view of that.

That's something I share with my PhD student Alex a lot more closely, in the sense that there is evidence that you can query the music data, so you can look at the digital signal and extract these features and come up with a relative quality level, but the human interface is always needed at the very top, and that's kind of already obtained in automated mixing. So, you can get to a rough mix very quickly without any human intervention and it will sound okay, but if you then want to start competing and getting really high quality sound, then you need some human intervention.

What we want to do now, and I'm not sure if Alex is going to have time to do this in his PhD, it might be a post-doc work, is to actually start using some more clever techniques like genetic algorithms to take those extra, let's say, 10–20% of the quality. So, we do automated mixing solely based on signal features and what we know from quality, but then the rest is a kind of much more genetic sort of evolution, and you'd try and see if one could apply those techniques to get to a much higher quality mix; that is comparable to a good producer, that's where we want to take it.

Q: Your point of view on that is very interesting. Have you heard of online automated mastering services? As you can imagine, it's become very controversial amongst mastering engineers, who won't understand the things that you understand about this. As you've just said, there are aspects of what a mix engineer does that you can automate, because there are base fundamental things that need to be done, and then human discretion needs to become involved. Would you be able to speak a little more about that, or whether you have an opinion that you'd be willing to venture, about the notion of automating mastering?

That's a discussion that has been had with the mixing process. When Josh and I started talking about automated mixing, it was faced with a lot of reaction from recording and mixing engineers saying "This is a ludicrous idea!" Of course, there is this danger of affecting people's jobs, and I think with mixing it is very clear that there are some aspects of the mixing that can be automated, especially in a live sound situation, for instance. You can do levels very quickly; you can do very simple compression very quickly, even to the point of dedicated compression to particular instruments. So you can detect whether it is a bass, and you apply the correct ratio, the correct gain thresholds, and so on. If it were a guitar, you'd apply something different. You might be doing that on presets, or you might be doing that on a much more symbiotic relationship between the channels, which is already possible. But of course on the top of that, then there would be the finer adjustments that the human needs to do. This, I think, would be a great tool.

However, with mastering, I would have to change my opinion just slightly, in the sense that yes, I agree that a large number of things could still be done automatically, so you could still look at the overall frequency range and see whether there are any imbalances or any discrepancies to a preferred curve, and you could do the same thing with dynamic range and loudness. Very quickly you could automate to get to a certain loudness, to a certain dynamic range, to a certain frequency range. Then, of course, the kind of "added art" could not come from a machine. It would need some kind of human intervention to actually make the most of the piece of music that's going to be released. Part of the mastering, some people could actually do in their own homes; they could kind of prepare it. Of course, the final touches, if you really want to do something professional, it has to go to a mastering studio, it has to go to a mastering engineer. Once it goes over that bridge, then, of course, the mastering engineer goes back and says, "I'd rather you didn't do anything, and let me do everything myself because I can get it to a better stage."

So automated systems of mastering, I think, are useful in getting the basics right. I think you can still put together a mastered version, but it won't be a professionally mastered version, it will be a kind of, "Ok, let's polish the corners here, according to some knowledge that we can automate, and deliver this." Online systems, doing this automatically, will be carrying out this process based on baseline principles, not refined adjustments. One could probably design an app or plugin to put on your DAW, which does more or less the same job, and I think there might be stuff out there that might already do that with presets and so on.

Q: It seems like a similar idea in that you can call up a preset and use that as a starting point.

There are other aspects to mastering, of course, so from a purely technical point of view, we could look at it as numbers and a response of the signal. One could use a preset and not even listen to it and trust that the preset is going to polish the corners and make things sound better. Mastering, just like mixing, just like recording, is as much a technical task as it is an artistic task: you need to know what to listen for, what can be done, and then you need to be able to hear that you can do that, added to having the tools to do it. So I don't think it's easy for someone, for instance, to master at home with bad acoustics, with perhaps not so good loudspeakers, and maybe not so good machinery, and of course they will also miss the experience that mastering engineers will have gathered over the years. That's the kind of aspect that you really can't automate, the added experience, the added knowledge, the kind of more symbiotic approach to each piece. An automated system can be quite blind to the signal that's coming in. Although in automated mixing we can already do some very clever dependencies on "if I've got these and these tracks, playing these and these notes, with these decays, then there's this amount of masking, so this other track needs to be EQ'd and placed at this level, with this amount of compression, for it to sound 'proper.'" That is possible, but of course there's a limit to what that can do, and I think in mastering I would say that the challenge there is, "Can a machine really take the best option? Can a machine make the most of what signal is coming in?" Perhaps not yet, because we still don't understand what features are consistently perceived as optimal quality.

Concluding Discussion

Russ Hepworth-Sawyer and Jay Hodgson interviewed by Jeremy Graham

Q: What inspired you to put together a book like this?

JAY: There is no single correct way to master a record. There are plenty of mistakes a mastering engineer can make—all of which qualify as "mistakes" because they either disfigure the initial premaster or they fail to gain client approval for some reason or other—but there is no single correct method for producing approved masters. Mastering is a wholly creative process, then. Everyone knows, or has easy access to information regarding, what numbers they need to peg their records to in terms of LUFs now; and they have a good sense of what sounds harsh or not. But how engineers get their records to reach those numbers and balances, and the various things they do to get them there, is an entirely personal process, and, again, it's wholly creative. Mastering isn't solving for x, in other words. As is borne out in all of the interviews in this book, even if it was solving for x, every engineer would be free to solve that equation however they see fit, so long as their masters meet with client approval. I wanted to hear how creative people actually mastered their records; I wanted to hear about mastering processes, that is, and not just some overarching ideal process existing only in academic ivory towers and armchairs. So I began work on this book with Russ because, for my part, I wanted to hear mastering engineers talk about their creative practice. Nothing more than that was at stake, for me.

RUSS: I've spent my whole professional career straddling between the professional world (engineering in studios, mixing and mastering) and the other side educating. For 20-odd years of teaching, any anecdotes about mastering, or even the "real" knowledge of those professionals, appeared not to be documented. We'd had the excellent *Behind the Glass* series from Howard Massey, but nothing really like it for mastering. So we saw a hole in the market, so to speak.

Q: What was the most challenging aspect of putting together this book?

JAY: Mastering is its own little fiefdom in the broader record production process. As such, many of the terms and techniques mastering engineers use and discuss are very particular and specialized. Finding a way to convey that specialized knowledge and skillset to a general interested audience, without intervening and changing or "dumbing-down" what the engineers actually said, struck me as the most difficult and challenging aspect of pulling this book together.

RUSS: I'd underestimated the time it would take. Getting the interviews was relatively easy. Recording them and hosting them took a bit of transatlantic synchronization, but again was easy. The transcription and subsequent authorizations take time, with so many professionals working at the top of their game. Jay and I kept deciding how we'd present the book to the world, either in one form or another. We settled on this . . .

Q: How do you think you can best help those individuals interested in audio and mastering?

RUSS: Education is key. However, despite my involvement in UK higher education part time, I note the disparity to the training I received to today. The opportunities to "learn" are different to the opportunities to "experience." The tape-op to engineer route (which as you'll hear from many in this book actually started in the cutting room) of mentoring has pretty much gone. Only the likes of Abbey Road can afford "runners" with degrees (many from Surrey University only) who are the equivalent, I suppose.

JAY: I think the only way I can help is through something like mentoring. I call it "shadowing." When someone is interested in mastering, and they leap the first hurdle of that process—which for me is simply coming up to me and saying they'd like to learn mastering and are willing to do whatever they need to in order to learn it—I invite them to "shadow" me. There is a couch at the back of studio, and they sit there for months, just watching me work. I invite them there for each and every session, no matter whether it's a big client or amateur. At the end of sessions, they are free to ask questions, but I invite them to just watch me work for a few weeks before even doing that. They are a shadow, in other words, and shadows don't make a sound. There's an element of ear training, there, which I think would be difficult to learn if the mentee became too caught up in audio concepts and psychoacoustics, et cetera. When they shadow, mentees will hear, over and again, premasters being balanced and brought to market levels. Through repetition, they become accustomed to the sound of a track going through the various processes it must before it is ready for release. More importantly, though, they see behind the curtain professionally. They watch client interactions, see me receive and prepare premasters and masters, and develop contacts for future work. They even see some of the promotional things we do, alongside more mundane activities like vacuuming the rug, dusting shelves, and so on.

Q: How important do you consider education to be as part of the audio industry, and could you explain why?

JAY: In my opinion, education is crucial. And not just for the information you learn in school, which is invaluable. What is truly crucial, for me, is immersing yourself in a network of likeminded individuals. We are all artists, in this craft, whether we never receive a single professional accolade over the course of our careers or we have so many Grammy awards that we use them as doorstops. And we do not live in a culture that is very kind to artists. We like celebrities, and we like successful people. I would even say we tend to worship them. But we don't reward creative lives *per se*.

Being with a group of people who, just like you, risked everything for a creative life, let alone to do the very kinds of creative things you want to do with your life, forces you to take your work seriously, I believe. You see that you are something like a "secret lineage holder," and that there are others just like you. You can learn something from all of them. Moreover, the network of students you will enter will help you achieve success in the future. I went to the Berklee College of Music, for instance, in the middle and late 1990s. Out of the 12-odd people who lived in the little corner of the second floor of the Commonwealth Avenue dormitory with me in the fall of 1995 and winter of 1996, during my first year there, some went on to win Grammies, some had number one Billboard hits, some received Platinum certifications, and so on. I wound up choosing a very different path, deciding to get a PhD and take a tenured position at a university here in Canada (I fell in love with my wife of 17 years and wanted to wake up in the same bed as her each and every day, and education seemed a way to maintain creativity without having to tour). But I stay in contact with many people from back then, and we share notes, and I learn more about the reality of the music industry through those contacts than I would anywhere else. This sort of experience is simply invaluable and, in my opinion, not available anywhere else than through education (online forums are a close approximation, but nowhere close enough to the "real thing," as it were, and they provide conceptual education rather than experiential, which in my experience simply covers the terrain of creativity with so many cognitive maps).

RUSS: The more books that can provide context and a historical account of the world of mastering, the better. The rest is knowledge (also can be gained from books or tuition in higher education) and experience gained by the apprentice route, or just doing. This book, I hope, will help those in search of that knowledge.

Q: How important is audio mastering for a project in terms of delivering a final product?

JAY: Crucial. Everything done during production is *subject to* mastering. Choices made in mastering affect every little choice made during production. It's a shame recording careers don't all begin with an apprenticeship phase in mastering. I think it'd be very eye opening to see the sorts of improvements, and damage, that can be done in mastering.

RUSS: Aha! Well this is key. There are shifting sands, presently. Ten years ago albums still (just) ruled the roost. As such, the product was a whole concept, a whole 60 minutes or whatever of programme. Think back to classic albums we grew up on, whether that be *Dark Side of the Moon* or *Thriller*, etc.—they all had what I call "flow." Flow is what binds the record together and ensures that the listener is "maintained" as they listen. The "perfection" of delivery, if you like. This is the mastering engineer's true skill (aside from the "making things pretty" element). If you can be transported away for the duration of that album without distraction, then you've done a good job.

Nowadays, the delivery is bitty. A single here, an EP here, an album whenever. The period between the first single and the fourth EP, the band has slightly changed, have recorded with five different engineers/studios, and then they ask you to pull it together. It's possible, of course, but it can be a "compilation" job rather than bringing birth to a concept.

Mastering is, however, under threat from many angles, as I'm sure we'll discuss . . .

Q: What's the most challenging aspect to mastering audio in your experience?

JAY: The answer to this question is dead simple, for me: developing the confidence needed to really determine where, if anywhere, premasters need to go to reach "deliverable" shape. This includes having the confidence to know when you don't need to do anything more, or when you can't do anything more because the premaster has been delivered in an unworkable shape. Oh, and having the determination to pursue and nourish your creativity, even when it seems the industry has rejected you or left you behind. I have done enough in my work that I can happily say I don't feel any need to prove anything to anyone, but there are still "dry" moments when the phone isn't ringing, and the email box isn't filling up with anything new, to make me question whether there's room for me still to continue working.

RUSS: As Jay says, it's the "knowing" that it's right. And the answer is "never." Then it's knowing you can do whatever is required with the tools you have to hand (rather than eternally looking for the über article). That comes with experience, knowledge, and confidence.

I've always chuckled when people say they are mastering but have inadequate monitors. The plugins today (I know colleagues disagree) are so good. DSP has come on so much, that I really don't think there's any absolute necessity to work out-of-the-box. However, one's D to A doesn't need to be a cheap one, and your amplification and speakers need to be the best you can afford. That's only half the battle. It's getting to know the speakers and the room inside out that takes investment—personal investment. Only with that investment comes confidence. Confidence means results.

Matt Colton said it for me in this book. To paraphrase, he said he knows what he needs to do to a track in the first minute. It's a gut reaction to what he hears, in his space, with his equipment. Yes, sure, like Mandy Parnell, Matt could master anywhere, probably on nearly anything, and do a better job than I, but he might take a little more time—it would not be "ninja" or subconscious. That's the power, skill, and so on, that mastering engineers bring.

Q: Out of everyone that was interviewed, were there any stand-out pieces of information, trends, advice, or recurring themes that were noted?

RUSS: A difficult one. Most colleagues said the same sorts of things, but the nuances were so different. The approaches different, the attitudes were shades of our profession. However, all were passionate about delivering high quality audio to the customer and public.

JAY: Actually, all of the interviews made a collective impression on me which I found striking, namely, that mastering is the fine art of figuring out a way to render masters from premasters. People do this in so many different ways that anyone insisting there is a "right" way, a "right" loudness, a "right" balance, and so on, usually comes off as sounding extremely amateur and insecure to me now. And this just makes sense: clear enough masters, work with enough clients, and you come to understand that mastering engineers take balances to a place that makes both themselves and their clients happy. Balances will always be dependent on input, then, which mastering engineers can't control—and the output mastering produces will always vary depending on where the client wants to take *their* art. I found that the more successful the engineer was, in terms of having their work applauded by their peers, the more open they were to different ways of working and to different aesthetic goals and priorities than what they might intuitively approach a project with. I admire the technician, for their devotion to learning, but they simply ply their trade according to an empirical disposition. They find comfort in numbers and equations and physics, all of which work to reassure them that their choices are correct. In my opinion, the best mastering engineers are artists who refuse to map their creativity with technical concepts and terms. These artists embody the old Zen adage that "not knowing is most intimate." Of course, both dispositions are fine. Whether through an empirical outlook, or an artistic one, the successful mastering engineer gains client approval for their work. That's it. No one way is better. To insist otherwise is like saying Tony Iommi sucked at guitar because he didn't play as fast or clean as, say, Yngwie Malmsteen. Both are great guitarists. They just do different things with their tools, and for different reasons.

Q: Was there anyone whom you couldn't reach that you would have liked to include in the book, and why? What would you ask them specifically?

JAY: There were plenty of engineers we couldn't reach, that we wanted to. Some just outright said no, we don't feel like doing interviews. If I could ask them a single question, it would be: why not? Is life really so busy? I don't think so. But then again, I'm not in their position. I'm here not because my livelihood depends on it but because I'm totally passionate about this work, and I want to continue to do it until I cannot physically make it into work anymore. So maybe it's like contacting a lawyer for an interview about the law, for some of these people—they just want to get away from work for a while, and we were just one more bother. But I don't know. I'm genuinely curious. I think the answer to that question would say a lot about what mastering is to these people, and what they think it ought to be.

RUSS: As Jay says, there were many who we reached out to who did not get back to us. There was one whole, large, cohort of engineers who promised to be involved, but their management then prevented them from getting involved. So if you see any gaps in the roster you'd expect, it's probably because of this. However, at this juncture, I think it is most incumbent for me to say how fortunate we were to have interviewed John Dent so shortly before his untimely passing. John left us too soon, but was a true pioneer in the field.

Q: Are there one or two questions that you wish you could have asked the interviewees, if you were to write the book again, or had more time?

JAY: To be honest, no. Those engineers who granted an interview were so tremendously generous with their time, and in sharing their insights and knowledge, that I wouldn't have felt comfortable asking for anything more. There's always going to be more work to be done at the end of a project like this. But I think we have a fuller picture of mastering now, having heard from so many mastering engineers about what they do and how they do it.

RUSS: There are probably many, but when you actually sit down to do a book like this, those questions become silly. What does the reader really want to know that they can't find out elsewhere (i.e., an engineer's view on something)? Once you start on this premise, then the questions narrow to what they became.

Q: What's some advice you could give to someone who's wanting to put together a book relating to audio?

RUSS: I've done many of them, and now it's hard finding something unique to bring to the table. The first books are difficult to improve upon at times. Jay is right—the money in publishing has dropped out, so you have to want to do it because you're passionate about the topic—as we are.

JAY: Make sure you love what you're writing about. Nobody does audio work to get rich, I don't think. If they do, I'd like to hear from them and spend some time with them on their yachts. In general, you are going to enter a world populated entirely by people who love the work and subject so much that they sacrificed social acceptance and mobility (after a certain age) for their craft. If you don't share that love and passion, they'll sniff it out in a second.

Q: Someone's just starting to get interested in mastering as a profession: name two pieces of advice to give them, one being something that they should definitely do, and one that they definitely should not do?

RUSS: Difficult. DO spend all your money on a decent mastering DAW package (not plugins at first), a decent D/A and excellent monitors and acoustic treatment for your room. If you are going analog, then some good outboard. Know everything you have well—really well!

Don't get sucked into the "grass is greener" on the other side with plugins. Once you truly know your stock DAW plugins can't do what you want, and it's not because a lack of knowledge, then go and spend. Know your compression and EQ back to front before pretending the latest $500 plugin will do it better. Your ears need to know!

Put it this way. If you'd spent $5,000 on an analog compressor, $4,000 on an analog EQ, you'd exhaust every ounce out of them before spending the money again on a different one (see Denis Blackham's experienced kit list). Your average mastering DAW has more than these two units—learn them!

JAY: This is a great question. I recommend you do two things (don't listen to people telling you what not to do): I would suggest getting very comfortable with receiving, and acting on, client revisions. There is an emotional down, even anxiety, we can feel when we receive notes early on in our careers. You do your very best work, agonize over the product, and you typically receive revision notes that, whatever the justification, boil down to: "I don't like it." Often, when you're early in your career, and working with other people early in their careers, too, those revisions come from the client not fully understanding what mastering is. They can expect you to fix things that can't really be fixed through mastering, in other words. So you're actually getting mix notes on your masters. This can be incredibly frustrating, especially when you haven't quite reached the point where even you understand this to begin with. So get good at communicating with clients, not panicking when you get unreasonable revisions, and seeing the whole project through to completion so everyone has a smile on their face at the end. I also recommend, early on, just hustling. Get as much work done as possible. Do it for free, or (better) for meagre pay, until you feel comfortable charging a sustainable wage. Grind out master after master, all day, every day, until you're sick of it and need a little break. Don't take the break. At least, don't take the break until you can afford to. Just keep going. This is a game of outlast, a marathon, not a sprint.

Index

Note: Page numbers in italics indicate figures.